LOCHEND

Monster Hunting
on the Run

Chasing Nessie During the
"Golden Age of Monster Hunting"

Published by
Joseph W. Zarzynski
P.O. Box 2134, Wilton, New York 12831

First Edition Paperback 2021

ISBN: 978-1-66781-050-8
Library of Congress Control Number: 2021919474

Front and back cover photographs (Loch Ness photographs by Garry Kozak, runner photograph from the Joseph W. Zarzynski Collection)

Page iv photograph of Urquhart Bay by Martin Klein

Photographs by the author unless otherwise noted.

Design: Toelke Associates, www.toelkeassociates.com
Printed in the USA
10 9 8 7 6 5 4 3 2 1

LOCHEND

Monster Hunting
on the Run

Chasing Nessie During the
"Golden Age of Monster Hunting"

Joseph W. Zarzynski

Printed by BookBaby

Dedication

For Tim Dinsdale, J. Richard Greenwell,

Marty Klein, Dr. Roy P. Mackal,

Ivor Newby, Loren Coleman, and

Tony Healy, who enriched our fund

of human knowledge during their

searches for hidden animals.

Also, to my darling wife

Mary Pat Meaney, for always supporting

my cryptozoological endeavors.

Contents

List of Illustrations

Preface

· ·

All right, I confess. For 17 years I was an unabashed monster hunter. I fervently chased denizens of the deep at waterways like Loch Ness and Lake Champlain, as well as at other so-called monster dwellings in Scotland. I was also at that time a diehard runner, what some people might call just your "average Joe marathoner."

During the early 1980s, marathon (26.2 miles) running had really caught on in the United States. I was one of that horde of distance-running devotees. I had taken up training for the sport at the age of 30, several years after I began my jogging regimen. Then, in May 1981 I completed my first marathon.

I'm not sure why I joined the hundreds of thousands of Americans who became marathon athletes. Maybe it was the trendy headband that did a decent job keeping sweat out of one's eyes while running. Or it could have been to meet other friends of jogging. Possibly it was because marathon running was considered the ultimate in fitness training. Whatever the reason(s), it just seemed like running marathons during the 1980s was the "in thing" to do for baby boomers.

Besides, 1984 was the year of the Games of the XXIII Olympiad, more commonly known as the 1984 summer Olympics, held in Los Angeles, California. You might say, during the summertime of 1984, all types of sports

P.1 Joseph W. Zarzynski running in a 13.1-mile race in 1981, part of his marathon training. (Credit: Joseph W. Zarzynski Collection)

were "in the air." One-hundred-and-forty countries participated in the 1984 summer Olympics, minus the Soviet Union and most other "Eastern Bloc" nations that decided to boycott the games. The communist countries' absence was in retaliation for the United States and many other pro-Western nations declining to attend the 1980 summer Olympics in Moscow, a formal protest of the USSR's invasion of Afghanistan in 1979.

By 1984, after finishing several marathons and an ultramarathon (any distance over the 26.2 miles) I was ready for a new-athletic endeavor. At 34 years old, I combined two passions—pursuing the lake monsters Nessie and Champ (Lake Champlain's Nessie-like animals) and ultra-distance running. On August 22, 1984, I undertook a solo run of the full length of Loch Ness. I chose to do the one-person ultra of Loch Ness, running from the village of Fort Augustus, located at the south end of the waterway, to the settlement of Lochend, at the very north of the loch. That ultramarathon was along a hilly course that totaled 28.5 miles. My solitary and what you might call somewhat-excessive run was also undertaken because I was a cryptozoologist (aka monster hunter),* a person who searches for "hidden animals." One of those bashful critters that I sought was Nessie, the "water horse" of Loch Ness. Nowadays, rather than being called "monsters," these evasive creatures are more often referred to as "cryptids," beasties whose very existence has been adamantly disputed by most in the scientific community.

Lochend—Monster Hunting on the Run, written over 2019–2020, is a celebration of the 35th anniversary of my one-person 1984 ultramarathon at Loch Ness. Moreover, my long run was undertaken toward the end of the "Golden Age of Monster Hunting at Loch Ness." That was a period when the believability in Nessie's existence by the American public was possibly at its greatest. In this tome, I describe details of my solitary jaunt, undertaken in 1984 through the quaint villages and hamlets of Fort

---

*The term "cryptozoology" has often been attributed to Dr. Bernard Heuvelmans, a renowned Belgian-French scientist, who specialized in the study of zoology. Two of his books, *On the Track of Unknown Animals* and *In the Wake of the Sea-Serpents*, had a significant influence on the development of cryptozoology. However, Heuvelmans acknowledged that he believed the terms cryptozoology (the science of studying "hidden animals") and cryptozoologist (a person that searches for "hidden animals") were probably coined by Ivan T. Sanderson, a Scottish-born and naturalized-American biologist and nature writer.

P.2 Map of the British Isles showing the location of Loch Ness. (Credit: Joseph W. Zarzynski)

Augustus, Invermoriston, Achnahannet, Strone, Drumnadrochit, and on past Lochend. The run was over the asphalt of busy route A82. In addition, I enlist the narrative of my ultramarathon to share the saga of the Loch Ness monsters* and also to relate anecdotes about those cryptozoologists who tenaciously worked to find an answer to this zoological thriller. After all, most people enjoy mysteries. The Loch Ness monsters, whether myth or reality, have for decades been one of the great enigmas in the world. Put on a pair of comfortable running or walking shoes, relax in your favorite chair, and read along as we chase these fleeting water critters.

---

*I use the term "Loch Ness monsters" because along with many other cryptozoologists, we believe Nessie is not a single animal, but a breeding colony. In 1978, Charles Wyckoff, a photographic expert with the Academy of Applied Science (AAS), theorized the Nessie community might total 20 to 30 animals (Reuter 1978). If they do exist, these mystery animals probably measure 15 to 30 feet in length. Four-and-a-half decades ago, Nessie was given a formal name—*Nessiteras rhombopteryx*. This appellation was bestowed because underwater photographs of a Nessie were believed to have been collected by the AAS, whose leader, Dr. Robert Rines, claimed were diamond-shaped flipper images from 1972 and also head- and full-body photographs of a Nessie taken a few years later. Therefore, Rines and wildlife expert Sir Peter Scott decided those photographs were proof enough to give the scientific name—*Nessiteras rhombopteryx*—to the Loch Ness mystery animals. That name meant—the Ness wonder [monster] with the diamond-shaped fin (Bauer 1988:17, 25). This was announced in an article in the December 11, 1975 issue of the journal *Nature* (Scott and Rines 1975:466-468). However, by the mid-1980s, these underwater photographs were criticized by some academics as not being Nessie animals (Naone 2007).

Acknowledgments

F irst and foremost, I want to thank Pat Meaney (aka Mary Pat Meaney), my wife, for her unwavering support during my research and writing of this book. On August 22, 1984, Pat drove an automobile as I ran the full distance of Loch Ness. Her motor cruise along route A82 that morning was challenging. Pat had little experience driving on the left side of the road as she motored ahead to make periodic stops to pass out water to me as unfamiliar traffic powered down near her.

Also, I want to acknowledge Tim Dinsdale (1924-1987), an amazing Briton and friend. Tim was a veteran of World War II (1939-1945) who later became an aeronautical engineer. He lived with his family in Reading, England, yet Dinsdale spent much of his adulthood at Loch Ness, perusing the loch looking for Nessie. Tim Dinsdale was also an accomplished author, penning several books about the Loch Ness monsters. We frequently corresponded by mail and on several occasions during my visits to Britain, we met to "talk shop" about Nessie and Champ, and to shore-watch over the fabled waters of Loch Ness. Additionally, Tim was the person who most encouraged me to run Loch Ness, combining two of my favorite endeavors—lake monster hunting and jogging. I am not sure how really accomplished I was at either, but both appealed to my sense of purpose and adventure.

There are numerous people and groups that have been running buddies during my 45+ years of jogging. I still run six-to-seven days a week, and I enjoy competing in half marathons (13.1 miles). These running buddies are: Charlie Babcock, Bob Baker, Mark Bessen, Jim Carlson, Dan Collins, Bob DeMarco, Greg Erwin, Bruce Farley, Joe Favat, Dennis Fillmore, Caroline Fink, Peter Finley, Danny French, Dr. Ted Gallagher, Hudson-Mohawk Road Runners Club, Linda and Art Kranick, Charlie Kuenzel, Dan Kumlander, Bruce MacWatters, Jack Mangini, Jeff Many, Rich Martin, Bill May, Bryna May, Scott McCloud, Pat Meaney, Joe Miranda, Mike and Tammy Newhouse, John Orsini, John Provoncha, Rick Rosebrook, Jan Roth, Tony Ryan, Joe Sporko, Paul Stevens, Richie Tanchyk, Bob Toth, Mike Valentine, Mike Veitch, Joan Williams, and Andy York.

I would be remiss if I did not also recognize those who contributed to my Nessie research and fieldwork: Dr. Henry H. Bauer, Richard Beckwith, Janet and Colin Bord, Ronnie Bremner, J. Richard Greenwell, Tony Harmsworth, Tony Healy, Rip Hepple, Paul D. Herbert, Jim Hogan, David James, Marty Klein, Garry Kozak, Mike Macdonald, Dr. Roy P. Mackal, J.A. Menzies, J.D. Mills, Peter Milne, Ivor Newby, Dr. Robert Rines, Adrian Shine, David ("Wiggy") Williams, and Nick Witchell.

In addition, there are a number of people and organizations that were instrumental in my search for the Lake Champlain monsters: Bill Armstrong, Bob and Paul Bartholomew, Ken Bartowski, John Becker, Dr. Russell P. Bellico, John Bierman, Rod Canham, Vince Capone, Loren Coleman, Mayor Erastus Corning II (Albany, New York), Dan Couture, Richard Cowperthwait, Terry Crandall, Bill Curry, Chip D'Angelo, Bob DuBois, Don Fangboner, George Forgette, George Early, J. Richard Greenwell, Dennis Hall, Bruce G. Hallenbeck, Chris and Gary Heurich, Scott Hill, International Society of Cryptozoology, Jim Kennard, Marty Klein, Garry Kozak, Dr. Roy P. Mackal, Gary Mangiacopra, Sandra and Tony Mansi, Scott Mardis, Don Mayland, Pat Meaney, Alan Neigher, Dave Pitkin, Anne Platt, Jim Randesi, Dr. Phil Reines, Susan Schmidt, Richard Smith, Ted Straiton, Spencer Tulis, Ralph Veve, Knight and Walter Washburn, Dwight Whalen, and Dr. George Zug.

Likewise, I want to acknowledge another project at Loch Ness. In 1976, Martin Klein (Klein Associates, Inc. and the Academy of Applied Science), assisted by Charles Finkelstein, discovered a twin-engine military aircraft sunk in Loch Ness. One of the locals told them that a PBY Catalina aircraft had gone down in the waterway during World War II. So, Klein's team announced they had possibly found a PBY Catalina. However, in 1978, Klein returned to Loch Ness with colleagues Garry Kozak and Tom Cummings. They brought an advanced-side scan sonar and collected a clearer sonograph of the sunken plane. Cummings, a vintage-aircraft buff, commented, "That looks like a Wellington." The team gave the information and location of their sonar target to Professor Robin Holmes of Heriott-Watt University in Edinburgh. Holmes used a remotely operated vehicle (ROV) with a video camera to survey the plane. It turned out it was a combat-hardened twin-engine British Wellington bomber, one of only two planes still in existence of 11,461 Wellingtons manufactured. In 1985, Oceaneering International, Inc., with support from other organizations, raised the warplane. I attended that operation as a member of the Loch Ness Wellington Association, one of the entities involved in the craft's recovery, and also as a freelance journalist covering the event for *General Aviation News.*

There were many organizations and people that were involved in the 1985 Loch Ness Wellington bomber project: Loch Ness Wellington Association, Brooklands Museum, Oceaneering International, Inc., Vintage Aircraft and Flying Association, Heriot-Watt University, British Aerospace, J.W. Automarine, RAF Museum-Hendon, *FlyPast* magazine, Robin Holmes, Morag Barton, Norman ("Spud") Boorer, Alfie Lyden, and others.

In addition, I'd like to acknowledge the people who helped my career in underwater archaeology, a field I began in the mid-1980s. My thanks to my colleagues with the underwater archaeology team Bateaux Below—Dr. Russell P. Bellico, Bob Benway, Vince Capone, Terry Crandall, and John Farrell. Others that assisted my underwater archaeological endeavors: Dr. D.K. (Kathy) Abbass (*Land Tortoise Radeau Survey and the Rhode Island Marine Archaeology Project*), Adirondack Experience—The Museum on Blue Mountain Lake (ex Adirondack Museum), America the Beautiful Fund, Amanda Andreas, AngioDynamics, Bill Appling, Bill Armstrong, Sandy Arnold, Bob Baker, Dave Beck, Pete Benway, Ed Bethel, Dr. Sam Bowser, Ted Caldwell, Capitaland Scuba Center, Chris Carola, Carolynn Carpenter, Steve Cernak, Norm Channing, *The Chronicle*, Cooper's Cave Ale Company, Lee Coleman, Tim Cordell, Paul Cornell, Barbara Crandall, Dr. Kevin J. Crisman, Maddy Cucuteanu, Dale Currier, Clive Cussler, Chip D'Angelo, Darrin Fresh Water Institute (Dr. Chuck Boylen, David Diehl, Dr. Jeremy Farrell, Dr. Sandra Nierzwicki-Bauer, and John Wimbush), Dr. James Delgado, Diver's World, Kerry Dixon, Bob Doheny, Carl Dunn, John Earl, Karen Engelke, John Farrell, Jr., Fort William Henry (Kathy Flacke Muncil, Robert Flacke, Jr., Robert Flacke, Sr., and Melodie Viele), Fort Ticonderoga, Ken Fortier, The French & Indian War Society at Lake George, The Fund for Lake George, John Gardner, William Gates, Elinor (Mossop) Gottschalk, Kip Grant, Lisa and Tony Hall (*Lake George Mirror*), Hall's Boat Corporation, Helen V. Froehlich Foundation, Historical Society of the Town of Bolton, The History Press, John Hoagland, James Hood, Dale Jenks, Bill

Key, Dr. Alexey Khodjakov, Marty Klein (Klein Associates, Inc.), Dr. Mike Koonce, Garry Kozak, Charlie Kuenzel, Bill LaBarge, Lake George Arts Project (John Strong and Laura Von Rosk), Lake George Association, Lake George Historical Association (Grace MacDonald and Marilyn Mazzeo), Lake George Marine Equipment Company, Lake George Park Commission, Lake George Steamboat Company, Lake George Volunteer Fire Department, Lake George Watershed Coalition, John Lefner, Left Coast Press, Doug Leininger, Bob Leombruno, Laura Lee Linder, Maria Macri, Rich Martin, Dr. R. Duncan Mathewson III, Mattison Family, Mark Matucci, Don Mayland, Chris McGuirk, Charles M. McKinney III, Lohr McKinstry, Kendrick McMahan, John Meaney, Pat Meaney, Lisa Miller, Marisa Muratori, Museum of Underwater Archaeology, M-Z Information, New York State Department of Environmental Conservation (Tim Hendricks, LeRoy Rider, Bob Thompson, Chuck Vandrei, Tom Wahl, and Gary West), New York State Divers Association, New York State Museum (Catt Gagnon, Dr. John Hart, Scott Heydrick, Andrea Lain, Phil Lord, Dr. Jonathan Lothrop, Stephen Loughman, Dr. Michael Lucas, Dr. Joe Meany, Kristin O'Connell, Ralph Rataul, John Ray, Dr. Christina Rieth, Molly Scofield, Susan Winchell-Sweeney, and Brad Utter), New York State Office of Parks, Recreation and Historic Preservation, New York State Office of General Services (Al Bauder and John Carstens), New York State Department of State, One Day Signs, Scott Padeni, Gary Paine, Denny Pajak, Alex and James Parrott, Mark L. Peckham, Joe and Peter Pepe, Jerry Pepper, John Pepper, Michele Phillips, Dave Pitkin, Paul Post, Preservation League of New York State, Tom Rasbeck, John Ray, Steve Resler, Rich Morin's Professional Dive Centers, Wit Richmond, Tim Rowland, Dr. Tim Runyan, Rural New York Historic Preservation Grant Program, Saratoga Springs City School District (Karen Cavotta, Bill Cooper, Bill Snyder, Jeff Sova, and Preston Sweeney), *The Saratogian*, Linda Schmidt, James Sears, Jon Smith, Speakman's Company, Dr. Megan Springate, Dr. David Starbuck, Jack Sullivan, SUNY Press, Bruce Terrell, Town of Lake George, Dave Van Aken, Victory Sports, Village of Lake George, Waterfront Diving Center, Dave White (New York Sea Grant—Oswego), John Whitesel, Wiawaka Holiday House (Christine Dixon, Dan Fisher, Joe Wiley, and Meaghan Wilkins), Ralph Wilbanks, Brian Wilcox, Claudia Young, Gary Zaboly, Marian and Walter Zarzynski, Stan Zeccolo, and Dick Zielinski.

Furthermore, thanks to all those who contributed photographs and illustrations for this book. They are acknowledged in the image credits.

Finally, I wish to recognize those who helped during the research, writing, editing, and layout of this book: Loren Coleman, Shuna Colquhoun, Paul Cropper, Joe Favat, Sharon & Basil Gribbon, Tony Healy, Wendy Herlich, August Johnson, Marty Klein, Garry Kozak, Charlie Kuenzel, Pat Meaney, Peter Pepe, Jeff Schmidt, Ron Toelke, Barbara Kempler-Toelke, and Anthony Lovenheim Irwin (Toelke Associates), and Nicole Wolfe (Union Endicott Central School District).

Introduction

M y mindset for the 1984 run along legendary Loch Ness began a decade earlier, in 1974. That's when I first began my interest, perhaps even my obsession, with investigating water monsters. As previously stated, the mid-1970s was an era when the American public's acceptance for looking for the Loch Ness monsters was possibly at its pinnacle. When I first became intrigued with lake monster hunting, it was more or less a field in which non-professionals could grab a camera and binoculars, and journey to Scotland's Loch Ness. If lucky, during those years you might have been able to work alongside some university-trained scientists, researchers who dared to risk their academic credentials by stalking so-called mythical creatures.

Additionally, the late 1960s and early 1970s ushered in a new phase at Loch Ness for cryptozoology, that of deploying emerging technology in hope of solving

I.1 The ruins of Urquhart Castle on the shore of Loch Ness, Scotland. (Credit: Joseph W. Zarzynski)

this zoological conundrum. Some of this sophisticated equipment had been developed during the aftermath of the USS *Thresher* submarine disaster in 1963 (Navy History and Heritage Command 2019). When that US Navy nuclear sub unexpectedly sank into the deep crevices of the Atlantic Ocean, it created a national initiative for the oceanographic industry and United States military to develop better instrumentation for scanning the remotest-ocean depths. You might say that the 1963 *Thresher* submarine disaster did for oceanography what the Soviet Union's Sputnik launch in 1957 did for the "race for outer space." Out of the *Thresher* catastrophe, some of America's "best and brightest" began developing state-of-the-art underwater equipment.

Martin Klein (Klein Associates, Inc., Salem, New Hampshire) was a graduate of Massachusetts Institute of Technology (MIT). He was a product of the post-USS *Thresher* inventiveness to develop better sonar* and other deep-sea imaging gear. In the early 1970s, the electrical engineer and sonar manufacturer of Klein Associates began collaborating with the Academy of Applied Science (AAS), attempting to decipher the Nessie riddle. With offices in Massachusetts and New Hampshire, the AAS aspired to solve this mystery by employing state-of-the-art sonar, underwater cameras, hydrophones, and even manmade scents they hoped would attract the elusive animals during their underwater fieldwork at Loch Ness.

Following his sonar fieldwork at Loch Ness in 1971, Martin Klein came to three initial conclusions. First, his sonar detected "large moving objects" in the loch. Second, there was enough fish life in those waters to support "a large creature." And third, there were sizable-underwater ridges along the steep walls of Loch Ness that might "harbor large creatures." The latter geology was dubbed the "Klein Caves" (Klein 1971:35-36).

From this surge of technological advancement emerged a buzz term at Loch Ness that thoroughly mesmerized me. "The Average Plesiosaur" was a witticism coined by Martin Klein. In late 1976, during an interview with an Associated Press reporter, Klein described a sonograph image that Charles Finkelstein and he acquired a few months earlier during their 1976 expedition at Loch Ness. The idiom, "The Average Plesiosaur," was also described in the article "Sonar Serendipity

---

*Sonar is an acronym for **so**und **na**vigation **r**anging.

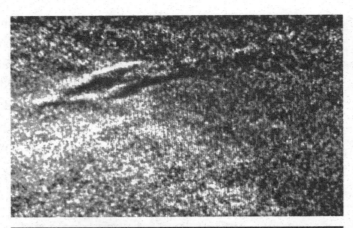

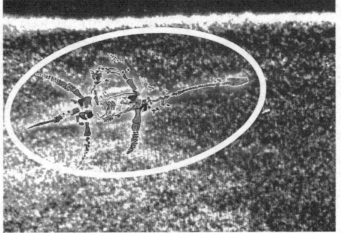

I.2 Martin Klein's side scan sonar image (top) from 1976 of the "The Average Plesiosaur," compared to a plesiosaur skeleton (below) superimposed over the sonar image. Klein teased a reporter that his team's sonar record might be a Nessie carcass lying on the bottomlands of Loch Ness. (Credit: Martin Klein)

in Loch Ness," penned by Klein and Finkelstein that appeared in *Technology Review* in December 1976. The prestigious magazine was published by MIT, one of our country's leading universities, located in Cambridge, Massachusetts (Klein and Finkelstein 1976:16).

Klein hinted that "The Average Plesiosaur" sonar image that his crew recorded at Loch Ness could be the carcass of a Nessie lying on the waterway's bottomlands. He believed the sonar target deserved further investigation. Reportedly, the anomaly was on the loch floor in a depth of about 330 feet (Klein and Finkelstein 1976:16). This peculiar object had a length estimated to be over 30 feet and somewhat resembled a plesiosaur, a marine reptile with a long neck, undersized head, four flippers (two-per-side), and a longish tail. The plesiosaur is a species that supposedly became extinct about 66 million years ago. Yet, each year there are several animal species, previously believed extinct, that are rediscovered by scientific investigators.

Klein was quoted in the 1976 Associated Press article: "We named the target 'The Average Plesiosaur' to tease our paleontologist friends. It will be interesting to find if the target is still there when we next go to look at it" (Associated Press 1976). This implied the sonar target could even have been of an animal that was alive and mobile, but had been lying on the loch bed at the time the sonar survey was undertaken.

I was captivated by the AAS's ultramodern surveying of Loch Ness, especially their sonar discoveries, and specifically, the sonograph dubbed—"The Average

I.3 An artist interpretation of a Klein side scan sonar such as that used in the search for Nessie in the 1970s. Note the Nessie at the left. (Credit: Martin Klein)

Plesiosaur." That summer I did not visit Loch Ness to look for Nessie, as I had done the previous year. It was my country's Bicentennial (1776–1976), a nationwide celebration of America's Declaration of Independence (July 4, 1776) from Britain during the American Revolution (1775–1783). Therefore, I decided to spend the summer of 1976 on the shores of another alleged-monster dwelling. My associate, John Bierman of Syracuse, New York, and I stationed ourselves at Port Henry, New York, gazing out over the 110-mile-long Lake Champlain. Port Henry is a hamlet in Essex County, New York that overlooks the lake, a waterway so mammoth, it was once briefly designated as the "Sixth Great Lake." Port Henry, at one time renowned for its iron ore mining, had a population of about 1,000 people. Back in the 1970s and 1980s, Lake Champlain was appropriately nicknamed "North America's Loch Ness," because of the many sightings of Champ recorded there over the decades.

John Bierman and I met at Ithaca College, located overlooking beautiful Cayuga Lake in western New York, when we were both undergraduate students. Our 1976 expedition was an attempt to film Champ, the Lake Champlain monster, or should I say, monsters, a breeding colony. John and I had no luck that hot summer trying to acquire film footage of a Champ, but we did stir up plenty of excitement in Port Henry, New York. After all, numerous residents of that area freely admitted that they had seen something strange and massive in the waters of Lake Champlain. To Port Henry residents and others, this animate oddity was known as Champ.

Several years later, in August 1984, and a few years after Marty Klein's seminal sonar investigations at Loch Ness, I was just a self-described "average marathoner." I likewise was a lake monster aficionado trying to photograph "The Average Plesiosaur." Thus, it seemed appropriate that I should run the length of Loch Ness. Remarkably, that athletic jog has the distinction of possibly being the first time* (see footnote next page) anyone ever ran the 28.5-mile-long distance along route A82—from Fort Augustus at the south end of Loch Ness, to past Lochend, at the north end of the waterway** (see footnote next page).

Years earlier, a journalist keenly observed the motivation of those people who undertook to prove the existence of Nessie. The reporter wrote in the July 4, 1961 issue of the *Daily Express* newspaper (London, United Kingdom): "Nobody will ever understand Loch Ness. Its conquest will be a greater triumph than the conquest of the moon" (Witchell 1974:13).

I.4 A painting of a Champ-like creature on a Port Henry, New York store-front window in the 1980s. (Credit: Joseph W. Zarzynski)

Additionally, there is an aphorism often tossed about by fanatical joggers—"Runners run." That catchphrase has been a mantra of mine. My jogging has mostly been done in the confines of the Saratoga Spa State Park, a recreation and cultural area in Saratoga Springs, New York. The stately grounds and resplendent buildings, just a few miles from my residence, are also listed as a National Historic Landmark (Breyer 1985).

---

*I sometimes wonder if the Scottish-ultrarunner Don Ritchie (1944-2018), certainly one of the greats in his sport, might have run route A82 along Loch Ness during one of his training sessions.

**This distance was measured in 1984 using a car odometer. The running distance included some side excursions I took into the parking-area stops along A82, to take periodic water breaks. I started and finished the ultrarun at locations that insured I covered the full distance of Loch Ness. The waterway's length, as "the crow flies," is over 22.5 miles, with some sources measuring it at 23 to 24 miles. The meandering roadway weaves along the shoreline and at times pulls a significant distance from the water.

Although basketball was my first sport of interest when I was a youngster, chiefly because I reached nearly 6 feet, 6 inches in height, jogging was the exercise regimen that later consumed me and what became a lifelong endeavor. So, in the 1980s, lake monster searching and distance running were two key components of my life. It was inevitable that I would combine the two and take a shot at running the length of Loch Ness. Hence, the title of this book: *Lochend—Monster Hunting on the Run.*

Chapter 1

Loch Ness—the Scene

M̲ost Americans and probably over half the world's population know at least a little about Loch Ness and its bewildering creatures. The acclaimed waterway is in Scotland in the British Isles. For the most part, the Loch Ness locale has cool summers and relatively-mild winters. Due to its vast volume of water, the loch generally does not freeze over in the winter. Loch Ness is an oligotrophic lake, that

1.1 Photograph of Loch Ness, Scotland in 1979, with the author conducting shore watching for Nessie. (Credit: Tony Healy)

is, one with "little nourishment." Thus, it has low-algal growth. Yet, because of the peat particles suspended in the water column, Loch Ness is not a crystal-clear lake. Rather, it has somewhat mediocre-underwater transparency (Mackal 1976:8). National Geographic Society underwater photographer Emory Kristof once said after scuba diving* in Loch Ness in the mid-1970s, that in its peat-stained waters at 40 feet and without a light for illumination was "like being lost in a coal mine" (Ellis 1977:775).

Loch Ness is part of Scotland's Great Glen. This impressive "rift valley" in the Scottish Highlands measures about 62 miles in length. The Great Glen runs from Fort William,** a town on Loch Linnhe, a sea loch on the west coast of Scotland, toward the northeast. The Great Glen ends at Inverness, a city located on the Moray Firth and an extension of the North Sea.

The Great Glen consists of several-natural waterways. These waters are from northeast to southwest: River Ness, Loch Dochfour, Loch Ness, River Oich, Loch Oich, Loch Lochy, River Lochy, and Loch Linnhe. Toward the northern end of the Great Glen lies Loch Ness, Scotland's second-largest freshwater body by surface area and its largest loch by water volume. Loch Ness is only second in surface area to Loch Lomand.

Loch Ness measures over 22 ½ miles in length with a width a little over 1 ½ miles. The loch reportedly has a maximum depth of approximately 755 feet. Earlier on in the twentieth century, the waterway was thought by some people to have a maximum depth of about 975 feet. However, modern-bathymetric instrumentation corrected that to 755 feet (Witchell 1974:19-20). Loch Ness is home to a variety of fish life including Atlantic salmon, brown trout, Arctic charr, pike, eels, minnows, sticklebacks, and other-pan fish (Harmsworth 2015). Salmon and trout are game fish eagerly sought by rod-and-reel fishing fans.

On the east side of Loch Ness*** (see footnote next page) lies General Wade's Military Road. This English military-built thoroughfare was constructed over sev-

---

*Scuba stands for self-contained underwater breathing apparatus.

**Near Fort William is the mountain peak of Ben Nevis, nicknamed the "mountain with its head in the clouds." Each year about 125,000 people climb the mountain peak with its elevation of 4,412 feet (Visit Scotland:2019a).

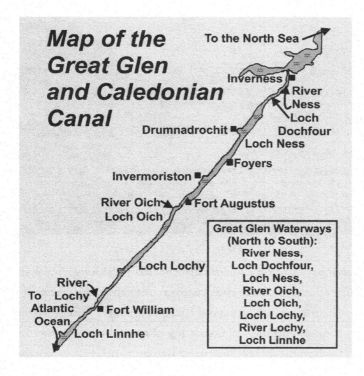

Map of the Great Glen and Caledonian Canal

To the North Sea

Inverness

River Ness

Loch Dochfour

Drumnadrochit

Loch Ness

Foyers

Invermoriston

River Oich

Fort Augustus

Loch Oich

Loch Lochy

River Lochy

To Atlantic Ocean

Fort William

Loch Linnhe

Great Glen Waterways
(North to South):
River Ness,
Loch Dochfour,
Loch Ness,
River Oich,
Loch Oich,
Loch Lochy,
River Lochy,
Loch Linnhe

1.2 Map of Scotland's Great Glen and Caledonian Canal. (Credit: Joseph W. Zarzynski)

eral years. It was begun nearly a decade after the Jacobite uprising in 1715, when James VIII attempted to regain the thrones of England, Scotland, and Ireland. From the mid-1970s to the mid-1980s, when I was conducting annual expeditions to Loch Ness, General Wade's Military Road was mostly a single lane. Every several-hundred yards or so there was a lay by for a vehicle to pull over to allow an oncoming car or lorry (truck) to pass.

Along the west side of Loch Ness is route A82, a panoramic road of two lanes that was built in the early 1930s. Its narrowness makes A82 a dangerous ride for non-vigilant drivers who may be gazing out over the water hoping to glimpse one of the water horses (aka Nessie). You could certainly make the case that route A82 opened the region to non-residents and thus helped spread the word about sightings of the mystery animals.

***Due to the centerline orientation of Loch Ness, some might refer to the east side of the loch as the south side and the west side of the waterway as the north side.

1.3 Route A82 along the west shore of Loch Ness in the 1930s. (Credit: Valentines of Dundee)

An early transportation route, the Caledonian Canal (60 miles long), is solely for watercraft. The canal opened in 1822; construction started in 1803. The man-made passage with its 29 locks, travels along the Great Glen with 23 miles of the canal being artificial channels and the rest consisting of natural waterways. The canal is not adequate for oceanic liners. Today the Caledonian Canal is mostly used by recreational- and coastal-fishing vessels (*Encyclopædia Britannica* 2019).

There are several municipalities nestled around Loch Ness. The largest of these, the city of Inverness, is just to the north of Loch Ness. Inverness is located where the River Ness, six miles long, meets the Moray Firth, the latter an inlet of the North

1.4 Boats in 1981, using the Caledonian Canal at Fort Augustus. Loch Ness is in the background. (Credit: Joseph W. Zarzynski)

1.5 Fort Augustus Abbey in the 1970s. (Credit: Joseph W. Zarzynski)

Sea. Inverness is regarded as the capital of the Scottish Highlands, and it is a customary stop for many visitors to the region.

At the south end of Loch Ness lies Fort Augustus. When I was frequently traveling to Loch Ness, 35- to 45-years ago, the focal point and biggest-tourist attraction in this village was the Benedictine Order's Fort Augustus Abbey. The Catholic monastery was developed from a Hanoverian fort constructed there by the British military over 1729 to 1745. The stone abbey is situated on a narrow peninsula at the south end of the loch. During my trips to Loch Ness in the 1970s and 1980s, the abbey was open to the public via tours (Fort Augustus Abbey n.d.). The religious institution, however, closed over a quarter-of-a-century ago.

Several-years back, British Broadcasting Corporation (BBC) reporters investigated the abbey when former students at the religious institution bravely came forward declaring they had been beaten and sexually abused by some of the staff and others. This was a horrific nightmare for those young people, and also a sad legacy for the monastery (Daly 2013).

Nonetheless, the village of Fort Augustus is a resting stop for tourists and a place for meals and shopping for travelers going between Fort William and Inverness. Little did I know during my first visit to Loch Ness in 1975, that nearly a decade later I would start my one-person ultrarun of Loch Ness from the center of Fort Augustus.

Chapter 2

Paving the Way—Running

· ·

Though my first-sporting passion was playing recreational basketball at gyms in my hometown of Endicott, New York, and being a basketball player on school teams, I initially became intrigued with jogging in September 1974. I was 24 years old and had just started work as an educator. I am not sure why I decided to become a school teacher, but I grew up enjoying elementary and secondary

2.1 Joseph W. Zarzynski's 1968–1969 Union Endicott High School varsity basketball team. Author is second from left, back row. Before Zarzynski became a long-distance runner, his preferred sport was basketball. (Credit: Union Endicott Central School District)

schools. My parents always encouraged me to get good grades and to appreciate the institution of learning. They constantly reminded me that getting an education would lead to having a better life. In retrospect, they were right. My many years of schooling helped me have a fulfilling lifetime.

I was born in Johnson City, near Binghamton, in upstate New York. I was a premature baby weighing 4 pounds, 4 ounces and thus I was an incubator child for several weeks. Our family lived in nearby Endicott in Broome County. The Village of Endicott was known not only as the "birthplace" of International Business Machines (IBM), but also it had a considerable blue-collar populace mostly employed with Endicott-Johnson Corporation, a major-shoe manufacturer. I grew up in a two-family, three-story house. Dad was a restauranteur who worked long hours managing a banquet facility. He later held a similar position at an American Legion post. Mom was the intellectual of our family. In 1937, she graduated high school at the age of 16, and was the senior-class valedictorian at a Catholic school in Binghamton (*Binghamton Evening Press* 1937). She later graduated from the College of St. Rose in Albany, New York. My mother was briefly employed as a school teacher during the Second World War (1941–1945), when my father, a U.S. Army sergeant, was stationed at several-military bases around the country. In my elementary-school years, my mother processed payroll for a department store in Endicott. She later worked for the state labor department until her retirement.

In June 1969, I graduated from Union Endicott High School (Endicott, New York) and then attended undergraduate school at Ithaca College (Ithaca, New York), about an hour car drive from our family residence. In May 1973, I received a Bachelor of Arts degree in history. Following graduation from Ithaca College, I attended Binghamton University, just a few miles from my home. There I received a Master of Arts in Teaching degree in Social Sciences.

In the summer of 1974, I was hired to teach social studies, Grade 9—Afro-Asian Cultures and also Grade 8—American History to the End of the Civil War, in the Saratoga Springs City School District in Saratoga Springs, New York. Saratoga Springs is about 35-miles north of Albany, the state capital of New York. My new employment began the day after Labor Day in 1974. I was delighted to have gotten that teaching job since Saratoga Springs had long been a cultural gem of the region and thus was a much-desired place to work and live.

2.2 Map of New York State. (Credit: Joseph W. Zarzynski)

Within a few days of beginning my teaching career, I met Charles Kuenzel. Charlie, as he was known to friends and colleagues, had also just been hired as an instructor in the same building where I taught. Charlie Kuenzel would go on to become one of the most talented science instructors in the state. A native "Saratogian," Charlie certainly knew the roads and byways of his hometown. Additionally, we only lived four blocks from one another, so he and I sometimes would hang out during our free time. I remember quite well, our first jog, undertaken one day after work in September 1974. Charlie had run cross country while in high school in the Saratoga Springs City School District. Therefore, he obviously knew more about competitive running than I did. As a local, he was likewise well acquainted with the neighborhoods around the "Spa City" and places to safely jog where traffic was minimal.

One late afternoon in 1974, I donned my pair of low-cut Converse sneakers. That foot attire, the Converse "Chuck Taylor" All Stars, was basically footwear for basketball, but it was all I had for any sports activity. We decided to go out for a short jog. Starting from Charlie's apartment, down the road we galloped. We negotiated a couple of turns at intersections and headed into the grounds of Skidmore College, located on a rolling hillside in Saratoga Springs. The private-liberal arts institution was founded in 1903 as an all-women's college. Originally its buildings and dormitories were located downtown (Skidmore College 2020a). In the early

1960s, the campus relocated to its current address on North Broadway. In the early 1970s, Skidmore College became co-educational (Skidmore College 2020b). After we entered the backside of the campus, we completed a circuit around the main road, a loop, and then we departed the college grounds, retracing our route back to Charlie's apartment. Though I thought I was in pretty-good shape having played basketball six-to-seven days a week for the past decade, it quickly became apparent that an intense jog was far different than running up-and-down the hardwood of a basketball court. Charlie was fast and our pace was not like basketball's intense-running bursts, jumping for rebounds, lateral movements playing defense, and all that interspersed with short rest. Rather, distance running was a continuous movement, repeated over-and-over for miles. In addition, I would soon often hear from fellow joggers, "the pace makes the race." I later spent weeks learning how to pace my jogging, developing a suitable-running gate, and learning how to properly breathe. On that first day of jogging in September 1974, Charlie ran a steady speed and I must admit, I was quite tired at the end of our 3 ¼ miles of running. That was my baptism into jogging and a distance that was far greater than the length of a high school or college basketball court that is only 84 feet.

Also, in 1974, running footwear had not yet evolved into the dynamically-designed running shoes of today. Charlie related to me how each runner on his cross country team received a new pair of Adidas running shoes, purchased by the school system. That company was one of the pioneers in the design, production, and evolution of athletic footwear. At that time, other sporting-goods corporations like Nike and Brooks were also beginning to manufacture shoes for jogging grunts.

Additionally, that was the era of such elite runners as Steve Prefontaine, a charismatic-Oregon University track star, Frank Shorter, the dynamic-1972 Olympic gold medal marathon winner, and an emerging-marathon great from New England—Bill Rogers. Like baseball, basketball, and football, distance running had its sport icons, too. Just a short time after my first-distance jog, I was gradually being converted to that sporting endeavor that would become a lifelong-exercise routine for me. What's more, I quickly discovered that during distance runs, one's mind often turned into a wonderful-thinking machine, able to devise remarkable strategies that helped one cope with the daily challenges of life.

Chapter 3:

The Loch Ness Monsters Emerge

. . . .

The construction of route A82, the two-lane road that I traversed on August 22, 1984 during my Loch Ness ultrarun, played an integral role in the emergence of Nessie as a global-scientific phenomenon. During the height of the worldwide-economic downturn of the early 1930s, forest lands along the western shore of Loch Ness were cleared and immense-rock formations were blasted apart to permit work crews to construct the roadway. When it was completed, visitors and locals could then travel along the length of the loch. They then had incredible views of the panoramic surface of Loch Ness, vistas that were previously unavailable.

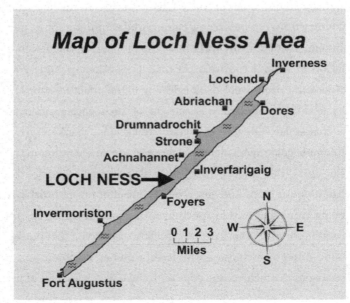

3.1 Map of Loch Ness Area. (Credit: Joseph W. Zarzynski)

Such was the backdrop for Mr. and Mrs. Aldie and John Mackay of Drumnadrochit. The village was on the west shore of the loch and had been settled around a bridge that crossed the River Enrick. The couple owned the Drumnadrochit Hotel. It was the afternoon of Friday, April 14, 1933, and the hoteliers were in their automobile driving home from a trip to Inverness. As their car was near Abriachan, several-miles north of Drumnadrochit, the couple had an unusual "encounter," one that would literally startle the world.

Aldie Mackay was the first to spot "it." Off to her left, about three-quarters of a mile away upon the surface of the loch was something strange, something that was immense and very unordinary. She yelled to her husband, "Look, John — what's that — out there?"

John Mackay gazed out over Loch Ness and then slammed his foot on the automobile-brake pedal. For quite a bit of time, about several minutes, the couple examined what they described as "an enormous animal rolling and plunging" on the water until it finally disappeared into the depths.

One might argue, well, the two owned an inn and tourism was down during what became known as the "Great Depression." The Mackays were later reported to not be "publicity hounds," that is, people who simply wanted to help fill their hotel. Their sighting of a massive-unidentified thing in the loch soon wove its way along the local

"grapevine" to Alex Campbell, the Fort Augustus water bailiff (aka game warden). Campbell's secondary job was that of a local correspondent for the *Inverness Courier* newspaper. He wrote a news story about this unimaginable incident (Witchell 1974:43-44).

3.2 Alex Campbell, the water bailiff at Loch Ness and an Inverness Courier *newspaper correspondent, shown here as an older man. His newspaper article in 1933 introduced the Loch Ness monster to the world. (Credit: Tony Healy)*

3.3 Dr. Evan Barron, editor of the Inverness Courier *newspaper in 1933. (Credit: Evaline Barron, Inverness Courier)*

The editor of this popular-Highlands newspaper at this time was Dr. Evan Barron. He reviewed Alex Campbell's article and reportedly commented, "Well, if it is as big as Campbell says it is we can't just call it a creature, it must be a real monster" (Witchell 1974:44). On May 2, 1933, two-and-a-half weeks after the Mackays' sighting on Loch Ness, the *Inverness Courier* published Alex Campbell's local-news account. In capital letters the newspaper story was entitled "STRANGE SPECTACLE ON LOCH NESS." Campbell's article reported that the "strange spectacle" was "no ordinary denizen of the depths," it was a "beast." Then the news flash called it a "monster." However, the media blurb about the Mackays' sighting went on to declare that several years earlier a group of Inverness anglers were in a "row-boat" crossing the waterway when they, too, "encountered an unknown creature" (Witchell 1974:45). The Fleet Street newspapers in London, considered at the time to be the very soul of journalism in the western world, ran with this riveting human-interest story out of Inverness. For the British Isles and the rest of the world, the Loch Ness monster was "born." A race was soon on for someone to photograph and solve this incredible-zoological mystery.

Chapter 4

Nessie's Cousin, Champ—
the Lake Champlain Monsters

Like many youngsters growing up, I was enthralled with dinosaurs of the Mesozoic Era (252 million to 66 million years ago). Later in my life, in September 1974, that childhood interest would unexpectedly be rekindled. It happened by accident and I suppose not too surprisingly, in a library. I was hunkered down at a research table in the Skidmore College library in Saratoga Springs, preparing lesson plans for my new job as a junior high school social studies instructor. The building was only a five-minute drive from my tiny-studio apartment and I enjoyed using the research facility. It had a collection of scholarly books, magazines, and journals, and there were several-photocopying machines, too. It was a weekend and thus the reference center was relatively quiet with hardly anyone inside the library. Following a couple of hours of grading student papers and writing lesson plans, I stood up from the table where I had positioned myself, to stretch my long legs. I began to take a leisurely stroll along the media center aisles hoping a brief walk would clear the mental cobwebs that had sprouted up in my brain. I sauntered down one of the passageways stopping occasionally to glance at some of the publication titles that were neatly shelved on the sturdy bookcases. Suddenly, out of my peripheral vision, I noticed the spine of a tome with a turquoise-colored book jacket. The title's text was in reddish ink. I pulled the book from the shelf and perused the cover's title. Set all in capital letters, it read—*MONSTER HUNT*. The author was Tim Dinsdale, identified on the front cover as the Director of Photography of the Loch Ness Phenomena Investigation Bureau.

4.1 The front cover of Tim Dinsdale's acclaimed-book Monster Hunt. *(Credit: Tim Dinsdale)*

I had been jolted as if stung by an ornery hornet. My senses were swiftly heightened. This eye-catching title and the vibrant-front cover were hypnotic. That hardcover publication, with its price listed at $6.95, would have an immediate and dramatic impact upon my life.

As a teacher in the municipality of Saratoga Springs, Skidmore College's library graciously permitted school-district instructors to sign out books. So, I took out the publication for the duration of a few weeks. The book literally created a new avocation for me—lake monster hunter.

Monster Hunt (Acropolis Books Ltd., Washington, DC, 1972) was a 295-page book that examined the evidence for Nessie, as well as being a primer on unexplained-serpentine animals reported to be secreted away in waterways in Canada, Ireland, Sweden, Russia, Malaysia, South America, and even the USA. The volume was also filled with compelling photographs, illustrative maps, detailed drawings, fascinating images, and remarkable-eyewitness sketches. Also, Dinsdale's *Monster Hunt* had a foreword written by Robert H. Rines, an attorney and the President of the Academy of Applied Science.

After delving into those pages, I soon found out that the word—"monster"—is derived from the Latin word "monstrum," meaning a supernatural being that is huge. I suppose a 25-foot-long sinuous animal with a small head, lengthy neck, burly body, and four flippers or appendages, what many eyewitnesses described as what Nessie looks like, could certainly be called a monster. The term—monster— was one I soon freely employed to describe the jumbo-sized unidentified animals said to be inhabiting Loch Ness and several other lakes around the planet.

Tim Dinsdale's *Monster Hunt* was originally published in 1966, but with another title—*The Leviathans* (Routledge & Kegan Paul Ltd.). A few years later the publication was updated by the author for a second printing in 1972, and it was given a more-sensational title—*Monster Hunt*. The *Leviathans* was not Dinsdale's first book on aquatic monsters. In 1961, Dinsdale's *Loch Ness Monster* was published by Routledge & Kegan Paul Ltd. That narration came out a year after he captured some spellbinding-film footage at Loch Ness of what he declared was a Nessie swimming on the loch's surface (Dinsdale 1961). Several editions of Tim Dinsdale's popular-1961 book, *Loch Ness Monster*, would later be published. Dinsdale dedicated his 1972 edition of the *Loch Ness Monster* book, "To those who search the 'Beastie' out" (Dinsdale 1972).

4.2 Monster hunter Tim Dinsdale in the field at Loch Ness. (Credit: Tim Dinsdale)

The year 1961, also saw the publication of another book about Nessie. That was *The Elusive Monster—An Analysis of the Evidence* from Loch Ness written by Dr. Maurice Burton, a British zoologist and popular-science writer. Burton's treatise was a counterpoint to the believers on the subject, as it was skeptical of the existence of the Loch Ness monster.

Dinsdale's and Burton's 1961 publications were not the first books about the Loch Ness monsters. Arguably one of the most popular of those early books on Nessie was written by R. T. Gould (aka Rupert T. Gould), a former-Royal Navy officer. His opus, *The Loch Ness Monster and Others*, was released in 1934.* Gould's book began:

In this book, I have tried to put on record what I believe to be reliable evidence concerning one of the most remarkable occurrences in recent years—the discovery, in a Scottish loch, of at least one specimen of what is probably the rarest and least-known of all living creatures (Gould 1976:ix).

Furthermore, Constance Whyte collected eyewitness sightings of the mystery animals of Loch Ness. She then authored a book about Nessie with the title—*More Than a Legend: The Story of the Loch Ness Monster* (Hamish Hamilton)—published in 1957. Thus, prior to Tim Dinsdale's books, there was considerable literature that delved into these perplexing creatures.

Why a liberal arts college library in upstate New York—Skidmore College—carried Dindale's *Monster Hunt* on its book shelves is to this day still somewhat puzzling to me. Nonetheless, finding the book was fortunate for me. It was a powerful "spacecraft" that launched me into a thrilling quest. In an era long before today, when information is readily available on the Internet at one's fingertips, I was engaged in reading and digesting Dinsdale's *Monster Hunt*. I gladly took in the riveting information provided on each page. I carried the publication to work, reading and even rereading chapters of special note. I studied it during lunch breaks in the

---

*In that same year, 1934, there were other books about Nessie. They were: *The Home of the Loch Ness Monster* by W. H. Lane (Grant & Murray) and *The Mysterious "Monster" of Loch Ness* by W. D. Hamilton and J. Hughes (Fort Augustus Abbey Press). The latter was a pamphlet published by abbey monks (Lochnessmystery.blogspot 2012).

4.3 Lake Champlain, the home of Champ, looking at the waterway from the Vermont side toward the New York shoreline. (Credit: Joseph W. Zarzynski)

school's faculty room as if, like my students, I might be quizzed on its contents. Then one day, a colleague, and like me, a member of the social studies department, commented on the book's title—*Monster Hunt*.

David Pitkin was a fellow teacher who certainly enjoyed anything that was "off the beaten path," stuff that was esoteric. Years later, after Dave retired from the classroom, he had a successful career writing about ghosts. But that autumn day in 1974, after Dave spotted the book title, *Monster Hunt*, he exclaimed, "Hey, Zarr, so you're interested in monster hunting for the Loch Ness monsters." His furry eyebrows arched and with a slight grin on his visage, my colleague suggested: "You don't have to go to Scotland to try to see lake monsters. Just take an hour-and-a-half drive north into Essex County to Lake Champlain. It's a huge and deep lake, one of the largest in North America. It also has reports of lake monsters."

And so began another facet of my life, looking for a "cousin" of Nessie. They were known as "Champ, the Lake Champlain monsters." What was even better, I didn't have to travel across the Atlantic Ocean to try to see one of these beasties. At the time I thought, what fantastic luck. The cosmos was certainly with me. It must be fate.

I should add that the water monsters of Lake Champlain also had a less frequently used nickname than Champ. New Yorkers who lived along the shores of Lake Champlain north of Essex, tended to call these cryptids—Champy. However, Vermonters and those residents of the Champlain Valley in New York state that resided in Essex and to the south, generally referred to them as Champ.

Chapter 5

Running Marathons and Ultramarathons

. . . .

As previously mentioned, basketball was my first-sporting enjoyment. I came from a family of some accomplished athletes, two in particular, who served as role models. One cousin, Robert Zarzynski, from my hometown, was an excellent basketball player at Endicott's Seton Catholic High School. At 5-feet-10, he would later shine as a guard playing basketball for Clark University in Worcester, Massachusetts. Robert was eventually elected into the Clark University Athletics Hall of Fame, a well-deserved recognition for his basketball accomplishments (Clark Athletics 2006).

Another superb athlete and relative was Thomas Linko, likewise from Endicott. Cousin Tom was exceptional at both baseball and football. At six feet in height, he was the starting quarterback on the Union Endicott High School football team. Tom, however, was most skilled as a baseball player, an outfielder and a fine slugger. After high school, he had a stint in the farm system of the New York Yankees. Tom didn't make the major leagues, but he clearly displayed his baseball prowess.

Both relatives inspired me to get into high school sports. With my height, over 6 feet, 5 inches, basketball obviously became my sport. As a senior in school, I became the starting center on the varsity basketball team at Union Endicott High School, but I was overshadowed by another teammate, Ken Nigh. Kenny was a gifted shooting guard who garnered all-league honors and later played varsity basketball at Ithaca College.

In my freshman year, I played basketball for only part of the season at Ithaca College. By mid-year, I retired due to a severe-ankle injury suffered the first week of practice. Though I had returned to the freshman squad following several weeks of physical therapy, it soon became apparent that I had an uphill road to being the basketball player I was earlier in the season. Also, I wanted to get my grades up as it was difficult for me as a freshman, adjusting to the rigors of both collegiate academics and sports. My mid-season departure from collegiate basketball actually worked out for me. My grades shot up and in 1973, I graduated college with academic honors.

On May 16, 1981, eight years after graduating from Ithaca College, I ran my first 26.2-mile race. It was the Champlain Valley Marathon in Clinton County, New York. The competition began at the Canada-USA border at Rouses Point, New York and the course meandered south along mostly rural roads, following the shoreline of Lake Champlain before finishing in Plattsburgh, New York. I can still recall the stiff headwind the runners faced for much of the event. It

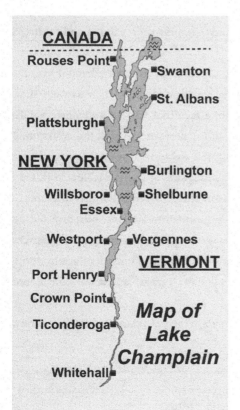

was a relatively warm day, too. It was no coincidence I selected this marathon. I would be running alongside the very waters I was investigating for its purported Lake Champlain monsters. I was excited to be competing in that marathon.

My time for the race was 3:56:11 (3 hours, 56 minutes, and 11 seconds). I ran the event with a teaching colleague from my school district, Greg Erwin, who pushed me in the early part of the marathon. I broke the four-hour barrier and I was certainly ecstatic about that. However, after that endurance run, like many first-time marathoners, I thought,

5.1 A map of Lake Champlain, known as "North America's Loch Ness." (Credit: Joseph W. Zarzynski)

S.2 Joseph W. Zarzynski in the 1981 Champlain Valley Marathon. (Credit: Joseph W. Zarzynski Collection)

hey, I could have done better. Therefore, it was on to further training and finding a second marathon.

Nineteen eighty-one, the year of my first marathon, was also a banner year for 26.2-mile races. On October 25th of that year, a pair of world records were shattered in the 12th New York City Marathon, a race that had 14,496 starters with 13,360 finishers. Alberto Salazar, a 23-year-old American, covered the course in 2 hours, 8 minutes, and 13 seconds, eclipsing the world record by 21 seconds; he won the race the previous year, too. On the other side of the 1981 New York City Marathon, 25-year-old New Zealander Allison Roe crossed the finish line first in the women's field with a time of 2 hours, 25 minutes, and 29 seconds, and setting a world record (Amdur 1981).* At the time, those distance-running performances inspired average runners like myself to get out, train, and complete a marathon. Finishing a marathon became an athletic "badge of courage," and it was considered to be a significant accomplishment.

I would run a total of ten marathons. Here are the year, race, and my race times:

---

*The 1981 New York City Marathon would later have some controversy. In 1985, it was determined the 1981 marathon course was actually a bit short (Burfoot 2016).

*1981 Champlain Valley Marathon (3:56:11),

*1981 Marine Corps Marathon (3:49:00),

*1982 Troy Heritage Trail Marathon (3:37:00),

*1982 Skylon International Marathon (3:35:57),

*1983 Foxtrotter Marathon (3:24:45),

*1983 Jim Thorpe Marathon (3:33:43),

*1983 Hudson-Mohawk River Marathon (3:27:42),

*1985 Atlantic City Marathon (4:00:21),

*1986 New York City Marathon (3:46:42), and

*2010 Rock 'n' Roll Las Vegas Marathon (5:26:01).

All of my marathon times except the 2010 marathon used what is called—"gun time." Gun time is from the beginning of the race when the starting gun is fired, until the runner crosses the finish line. Today, however, with timing chips embedded into runners' bibs, you have what is known as "net time." That is, race time does not begin when the starting gun goes off. Rather, it is the time from when the runner crosses the starting line to when the competitor ends the race crossing the finish line. If my first nine marathons were contested today using net time rather than gun time, all of those races would have been faster by many seconds, possibly in some cases, up to two minutes or more.

My personal record for a marathon, 3:24:45, would come on December 3, 1983, in Foxboro, Massachusetts at the Sixth Annual Foxtrotter Marathon. The race was sponsored by the Foxboro, Massachusetts Foxtrotters, a New England-running club (Foxboro Foxtrotters 1983). The event had a somewhat unusual marathon course as it repeated parts of the route. Thus, in that marathon, I was able to get the "lay of the land" after my first circuit. That helped me devise better strategy for my running pace later on in the competition.

Another race I completed was the 1983 Hudson-Mohawk River Marathon. This was an extremely flat course that started in Schenectady, New York. Runners headed east, mostly on a asphalt-bike path that paralleled the Mohawk River to the Hudson River, before turning south to finish in downtown Albany, New York.

I was pushed in that 1983 marathon by one of my running pals, Tony Ryan. He taught English in the Saratoga Springs City School District. Tony had been a high school-basketball athlete and he played collegiate basketball, too. He took up distance running in his early thirties. Tony was a funny person who could also recall newsworthy-sporting events that happened during his lifetime. Marathon training with Tony was pure pleasure as he regaled me with sports stories and amusing tales. He was a solid runner, too. Tony pushed me to be faster than I probably would have been, had I trained and run the race solo. I covered the 1983 Hudson-Mohawk River Marathon course in 3:27:42, just three minutes off my personal record. None of my marathon times were elite. I was in that middle echelon, "just average" for a marathon runner.

My most recent 26.2-mile race was over a decade ago. It was the December 2010 Rock 'n' Roll Las Vegas Marathon held in the heart of downtown Las Vegas, Nevada. I trained for almost a year for that race, probably because I would be competing in the event during an important-age milestone in my life, 60 years old. Also, I had not run a marathon in 24 years, so I was definitely not in tremendous-running shape. Thus, it took me quite some time to prepare for the race.

Other Rock 'n' Roll marathons were held around the world, too. A few of those sites were in: Washington, DC, Savannah, New Orleans, Seattle, and even some cities overseas (Rock 'n' Roll Marathon Series 2019). Additionally, in these running events, about every couple of miles, there was a musical band that entertained the runners and the throngs of supportive spectators.

I selected the Rock 'n' Roll Las Vegas Marathon mainly because our lovely goddaughter, Lisa Randesi, lived at that time in "Vegas." I thought this would be an ideal destination-marathon and an enjoyable spectator-event for both my goddaughter and wife. However, Lisa, having never witnessed a marathon competition, was somewhat aghast when she observed the slower marathoners toward the rear of the race pack. Some, like myself, were struggling quite a bit, barely able to keep going in sunny and warm Las Vegas.

The marathon was a rude awakening for me. After all, I was 60 years old and this was a marathon that was being played out in the hot desert of southern Nevada. Though early December, it was nonetheless warm and without a single cloud in the bright-blue skies hovering over "Sin City." Also, part of the race, much of the latter half, was over a gradual, but lengthy-hill climb from the Las Vegas Boulevard, known as "The Strip," heading up toward Red Rocks. The majestic-mountain range was situated just to the west of the city known for its slogan, "What happens in Vegas, stays in Vegas."

Additionally, the national-rodeo championships were in Las Vegas that same weekend. What a dichotomy, marathon runners and "cowboys and cowgirls." There were rodeo fans all over the place, too, watching in astonishment as they must have thought the marathoners were absolutely crazy to run in a desert. During the race, I frequently heard some of those bronco busters shout out to the runners, "Hey, you know you can get there faster on a horse." Indeed, they were correct with their sarcasm.

Well, I had an absolutely terrible race that day in the arid Nevada dustbowl. My time for finishing the 26.2-mile course was very disappointing—5 hours, 26 minutes, and 1 second. But, as my supportive wife, an enthusiastic cheerleader, reminded me over-and-over in the days following the marathon, "You finished, and you finished upright." She was right about me completing the marathon, but I must admit I was barely vertical when I ended that test of physical endurance in the desert.

I sometimes recall my struggle to reach the finish line, located in the parking lot of the Mandalay Bay Resort and Casino along the south end of "The Strip." Burned

5.3 Joseph W. Zarzynski running in 2010 Rock 'n' Roll Las Vegas Marathon. (Credit: Pat Meaney)

out, tired, dehydrated, and with a sore-lower back, I completed the long race just behind one of the running-Elvis impersonators. Each of those Presley doppel-gängers was dressed in a one-piece sequin-laced jumpsuit that would have been most appropriate for "the King of Rock and Roll." Some Elvis lookalikes that day even donned a black-head wig. In the case of the Elvis runner who was just strides ahead of me, he unbelievably carried a blow-up guitar, too. As we approached the finish line, it was a dash to the end—"Elvis versus me." That day, "Elvis" bettered me by several seconds.

Another of those Elvis Presley impersonators who had completed the marathon and who also was one of the few-dozen runners that got married on "The Strip" during the race, became a real-life hero later in the day. Dr. Claudio Palma, a San Francisco physician, was strolling by a hotel restaurant just after finishing the grueling race. Palma was still in running shoes and of course, he was dressed in Elvis attire. Suddenly, a woman in the eatery, a diabetic, collapsed. She was bleeding from her head and mouth from the fall and the lady reportedly had no pulse. The Elvis-looking anesthesiologist garnered enough energy to sprint over and start to perform CPR on the felled victim. The stricken woman finally opened her eyes, looked up, and later told a newspaper reporter that she wondered why Elvis was kissing her. Dr. Palma, the Elvis look-a-like, employed his medical expertise to revive the fallen person (Caulfield 2010). I wonder if Palma, after his timely recuperative intervention and while receiving congratulations from the growing crowd around the victim, uttered Presley's tagline, "Thank you...thank you very much."

At the finish line of the Rock 'n' Roll Las Vegas Marathon in 2010, I likewise was just several paces in front of a young duo. They had stopped for several minutes during the race at a run-thru chapel along the course. There they, too, received their nuptials. For the last-hundred yards I miraculously held off this darling, just-married couple. The joyful pair were holding hands and dragging the proverbial-aluminum cans tied on by strings, a common custom for those "just married." I suppose in the wedding tradition of "something old, something new, something borrowed and something blue," I was the "something old." I discovered the day after the race, that two of my toenails from the punishment of the long run were "something blue" black-and-blue that is, bruised from over 50,000 steps taken during the marathon. Yes, reality is sometimes sobering, especially in glitzy Las Vegas.

Back in the early 1980s and after already having participated in several marathons, I graduated to a new-running attraction or what a friend called, my next "abuse my body" compulsion, that of ultramarathons. As mentioned, these were running events of any distance farther than the 26.2 miles of a marathon. I was just an "average marathoner," so I thought, well, maybe I could be a better ultrarunner.

To prepare for those types of distance races, I immediately subscribed to *UltraRunning*, the leading magazine in the field. I quickly learned that the standard ultramarathons were 50 kilometers (31.1 miles), 50 miles, 100 kilometers (62.1 miles), 100 miles, and 24 hours. The 24-hour competition was how many miles one could run or run-walk in a single day.

Beginning three years after my Loch Ness ultrarun, I did the first of my two 24-hour races. Those two ultramarathons were held in Westport, New York. This pleasant-lakeside town is located on the western shores of Lake Champlain in Essex County. I had visited Westport on numerous occasions, launching boats from its boat ramp for my cryptozoological fieldwork to undertake side scan sonar excursions and to conduct scuba-diving operations. Furthermore, I had presented several lectures in that town on the topic of Champ, and even interviewed a few "Westporters" that had Champ sightings. Since Westport was only a little over an hour-and-a-half drive from where Pat and I lived, that would make the logistics of the 24-hour ultrarun rather easy. Therefore, I decided to enter the 13th Annual Essex County 24 Hour Ultra-Marathon in Westport, being held over July 11–12, 1987.

Unfortunately for the runners, race officials, event volunteers, and spectators, that day was sunny and temperatures reached nearly 90 degrees Fahrenheit. Race contestants opposed one another over a racing track at the Westport fairgrounds, a site that was a little less than a mile from the shores of beautiful Lake Champlain. I finished in 14th place out of 23 ultrarunners. I completed a total distance of 62 miles (Lopez 1987). The competition that day and night was won by an astounding ultrarunner who just happened to also be the world champion for that distance.

Yiannis Kouros was an international-running icon from Greece. He finished the 1987 Essex County 24 Hour Ultra-Marathon race having run a total of 142 miles (*Valley News* 1987:16-17). It was exhilarating for me, and I am sure for most of the other runners, to have competed on the same half-mile track with a world champion. The cinder track at the fairgrounds had been primarily utilized for years

5.4 Yiannis Kouros, a world-record holder of several running events, wearing bib number 101. Kouros is standing in front of the author (tall person, white shirt, and cap) at the start of a 24-hour ultrarun at Westport, New York in 1987. (Credit: Pat Meaney)

for harness racing during summer fairs. So, it was an ideal site for the 24-hour ultramarathon. The sporting event was coordinated by Dr. Robert Lopez, a Westport veterinarian and the local-running guru. Yiannis Kouros was so far ahead in the 1987 race, he just stopped running at 142 miles. Amazingly, there were still many minutes left in the ultramarathon. A decade later in 1997, he set the world record for 24 hours, an incredible 188.5 miles (Crockett 2019).

The following year, the 14th Annual Essex County 24 Hour Ultra-Marathon in Westport was on July 23-24, 1988. I again decided to enter the race. The weather that day was fortunately not as hot as the previous year. I significantly improved my mileage from the 1987 race. In the 1988 ultramarathon I finished 6th out of 23 ultrarunners. I covered a distance of 76.25 miles, an improvement of 14.25 miles from my 1987 race (*The Pace Setter* 1988). The experience of having run the event the year before, combined with cooler temperatures, certainly contributed to my vast improvement.

Another distance-running event I completed was a six-hour hill climb, known as the Overlook Overload in Woodstock, New York in the Catskills Mountains. The competition, a run up-and-down Overlook Mountain Trail, was on September 29, 1985. I entered the unconventional race with two other Saratoga Springs athletes, Paul Wanser and Arturo Santiago. The two acquaintances were from a local gym and they were not your average ultrarunners. Rather, the two were built more like football linebackers. Both were what you might call, "gym rats," guys who tossed around lots of iron weights to build thick muscles. Nonetheless, Paul and Artie were up for any unusual sporting challenge to test their athletic resolve.

The goal of the Overlook Overload race was how many times you could run up-and-down a mountainside. It was a flawless-autumn day with pleasant weather and the seasonal foliage was just beginning to change. It was an ideal setting for a distance run. I completed five laps of the challenging course plus an additional 1 ¼ miles. I was 7th out of 21 ultrarunners (Dick's Sports 1985), a finishing place I thought was quite respectable considering I had not done a heavy amount of hill training leading up to the competition.

Chapter 6

My Loch Ness Expeditions
Over the Years

. . . .

Over a 10-year period from 1975 to 1985, I made eight-fieldwork trips to Loch Ness. These I preferred to call "expeditions" because as I perceived it, I was traveling to Loch Ness to explore its waters and conduct scientific research. My Loch Ness expeditions were in 1975, 1977, 1978, 1979, 1981, 1982, 1984, and 1985. Here is a summary of a few of my expeditions that serve to give a flavor of what an American cryptozoologist did at Loch Ness during that era.

1975

My first expedition to Loch Ness was over the summer of 1975. I had just completed my first year as a social studies teacher in the Saratoga Springs City School District. I flew to London, England, and spent some time touring that historic city before taking a train north to Scotland.

I spent nearly two weeks in Scotland. Much of that time along the shoreside of Loch Ness conducting solo searching for Nessie. My shore watching consisted of monitoring the loch with binoculars and a camera. I saw no Nessie creature during my 1975 expedition, but I learned a lot about the Loch Ness environs, and I sure met many wonderful Scots. It was indeed a productive trip for a novice cryptozoologist and it prepared me for future expeditions to Loch Ness.

1977

My second expedition to the fabled shores of Loch Ness was in 1977. That trip, I was better prepared as I then knew where to go and whom to visit. My journey to the United Kingdom was from April 24 to May 1, a visit during my spring break from teaching in 1977. Since many students and families from northern states went on spring break to Florida and other warm climates, I preferred a non-traditional getaway. I journeyed to rainy and cool Scotland in April to search for what many people believed were phantom animals—Nessie. I had decided it was much more constructive for my overall studies in cryptozoology to go to Loch Ness in the springtime and then devote my summer vacation from teaching to undertake Champ fieldwork at "North America's Loch Ness."

On April 24, 1977, I flew to London from John F. Kennedy International Airport (aka JFK Airport) in Queens, New York City. From London, I caught a shuttle flight to Edinburgh, Scotland. I spent the night in the Scottish capital and took a morning train to Inverness arriving in the "Highlands" in the early afternoon. I was determined to make the rest of the day productive. I secured my automobile rental in Inverness and drove down route A82 to Fort Augustus. There, I briefly met with Father Gregory Brusey at the Benedictine Abby. The Catholic cleric had a notable sighting of a Nessie animal in the autumn of 1971 and I was interested to hear more about what he saw.

6.1 Nessie-eyewitness Father Gregory Brusey standing next to Joseph W. Zarzynski in 1979. (Credit: Tony Healy)

Father Brusey had recalled his sighting in Tim Dinsdale's book—*Loch Ness Monster* (second ed., 1972). Brusey's glimpse of a Nessie occurred on October 14, 1971. The cleric was with a friend, Roger Pugh, a church-choir director and organist from London. Brusey had taken Mr. Pugh down to the water's edge of the abbey grounds to show him the beauty of legendary Loch Ness. Father Brusey said it was a sunny day with calm waters. The two men walked out onto the stone jetty near the boat house and admired the then tranquil waters of Borlum Bay. Suddenly, the two spectators noticed a "terrific commotion" in the loch, at a range of 300 yards.

Brusey later wrote, "In the midst of this disturbance we saw quite distinctly the neck of the beast standing out of the water to what we calculated later to be a height of about 10 feet." They saw the aquatic behemoth for about 20 seconds, but the two eyewitnesses saw no humps (Dinsdale 1972:151-152). Father Brusey and I scheduled a time for us to have tea with some of the abbey's monks and to further discuss his 1971 sighting of Nessie creature.

Later during my first day at the loch in 1977, I took route A862 (aka General Wade's Military Road) and drove over to see Frank Searle. He was a controversial Nessie hunter whose waterside camp, dubbed "The Frank Searle Loch Ness Investigation," was located in Foyers, on the east side of the loch. Searle, an ex-soldier, first began his monster-hunting surveillance at Loch Ness in 1969 (Coleman 2005). After visiting with Frank Searle I wrote in my journal about his waterside residence: "Searle's camp consists of a wooden display shack of his Nessie photogs [sic] and various media articles. He lives with a small trailer as his home….Searle had a small cabin cruiser that has now replaced his dingy."

I had spent about 20 minutes talking to the monster-hunting enthusiast. He professed to have had 28 sightings and to have snapped "pictures on eight different occasions" (Searle 1977). Many of those images were claimed by his critics to be dubious photographs. Most of the time during our April 27, 1977 conversation, Searle spoke, giving me his perspective on how one should go about chasing these evasive animals. I recorded in my journal the main points of Searle's discourse to me:

1. Nessie is an animal that is fish-like and reptilian in shape.
2. Boats don't bother Nessie as she [his words] has become use to it.
3. When shore watching, it is better to pick a small field of view so as to get that close in, conclusive photograph rather than a photograph at a distance.
4. Searle said … there have been three sightings of Nessie so far this year [1977].
5. Searle said it was better to use a 16 mm [movie camera] rather than 8 mm for shore watching duty.

Of course, at the time, I had an 8 mm Nikon movie camera, but I didn't admit that to him. Frank Searle published a 30-page booklet in 1977 that he entitled *Around Loch Ness—A Handbook For Nessie Hunters.* He sold it at his little exhibit in Foyers (Searle 1977).

On April 28, 1977, I had tea with Father Brusey and his fellow monks in a spacious dining hall in the abbey. I then motored over to see Alex Campbell, the former water bailiff of Loch Ness who also was a part-time reporter for the *Inverness Courier.* I first met him during my 1975 visit. We had a short conversation as Campbell had to attend a funeral, but he did mention two surprising things to me. Campbell, who was the newspaper correspondent that first wrote about the Mackays' sighting in 1933, commented that he had seen a Nessie animal on 18 occasions. Further, Campbell told me he knew a woman who claimed she had "seen the beastie" 23 times. I suppose if you spent enough time looking at the loch, you would have a reasonable chance of having multiple sightings of Nessie. After leaving the residence of Alex Campbell, I wondered if I'd ever be lucky enough to have a sighting of the Loch Ness monster.

After several days in the United Kingdom, on May 1, 1977, I flew back to the USA. My monster-hunting trip to Loch Ness was brief, only several days, but it had been a busy—and unconventional—spring break. On May 2, when school started up after spring break, some teaching colleagues and students had noticeable tans from their exotic spring vacations to sunny Florida, South Carolina, and the Caribbean region. I was still somewhat chilled from the brisk spring weather of the Scottish Highlands. Yet, I was quite content because unlike them, I had pursued lake denizens in Scotland.

1978

In 1978, during another vacation from teaching, I flew to Britain to visit Loch Ness. This time I was accompanied by a friend, David Williams, known as "Wiggy" to his buddies. Williams grew up just a few miles from my hometown. In 1978, Wiggy resided in Maryland. He was an engineer at a manufacturing facility in greater-metropolitan Washington, DC.

During our week-long April 1978 field trip, we were amazingly fortunate to have observed an "unidentified swimming object" (USO) on the surface of Loch Ness. Thus, my wish to observe a Nessie animal at least once in my lifetime became a reality. It was April 25, 1978. Wiggy and I were driving along route B852 on the east side of the loch. It was late in the afternoon, approaching dusk, but there was plenty of light and visibility was excellent. We were traveling northeast, past Boleskine House (aka Manor of Boleskine) and the hamlet of Inverfarigaig.

Boleskine House was originally built in the mid-eighteenth century and its most unusual owner was the early-twentieth-century mountaineer and occultist Aleister Crowley. Jimmy Page of the band Led Zeppelin later owned the former Crowley property that overlooks Loch Ness.

The USO that David Williams and I spotted that late afternoon just before dusk appeared to be animate, moving along the water's surface heading southwest. It looked like a hefty dark hump going from near shore out into the deeper water of mid-loch. We observed no head and neck, nor a tail, or any appendages. I would estimate the exposed part of the mystery animal, that looked a bit like an overturned boat, had a length of 15 to 20 feet. By the time I stopped our sedan and backed up to get a better vantage point, the animate object had unfortunately disappeared into the depths. This was my one-and-only sighting of what might have been a Nessie animal.* (see footnote next page) I was 27 years of age. In retrospect, I was very lucky I did not drive the vehicle off the road as both my colleague and I were surprised, even somewhat shocked, at what we had just observed. Years later, David Wiggy Williams, fellow-Nessie eyewitness, was the best man at my wedding to Pat Meaney.

1981

On July 11, 1981, Pat drove me to JFK Airport in New York City. We had only been going together for a few months, so my expedition to Loch Ness that year was bittersweet as I would be away for nine days from my companion. After flying to Prestwick Airport in Scotland and checking through customs, I secured my rental car. Soon, I was cruising along on route A898 bound for route A82. The narrow roadway would take me past Loch Lomand and into the heart of the Scottish Highlands. The drive along route A82, though over a twisting roadway, was spectacular. I spent the night at Glengarry Castle Hotel at Loch Oich, a four-mile-long waterway that is part of Scotland's Great Glen.

The Glengarry Castle Hotel was formerly known as the Invergarry House. The lodging overlooks Loch Oich and nearby sits the impressive ruins of Invergarry Castle, an L-shaped stone fortification, that once served as the seat of the Clan MacDonnell of Glengarry. The first martial structure was erected in the seventeenth century and over the next century-and-a-half, the citadel was occupied by several different factions. It was said that Bonnie Prince Charlie even spent time there from 1745 to 1746 (Glengarry Castle Hotel 2020). I seized the opportunity and took a brief trek along the shores of Loch Oich, camera in hand to inspect the old fortress and to look for the reputed leviathan of the waterway.

Due to the time zones crossed and the lengthy air travel, when I awoke the next day, it was July 13. I ate a hearty breakfast and then checked out of the inn. I

---

*This was not my only sighting of an USO. I also once had a sighting of Champ. It was August 10, 1988, at 6:50 p.m. It was sunny with superb visibility. Several people, including my wife Pat Meaney and myself, observed a strange animal that we believe was one of the Lake Champlain monsters. We were conducting side scan sonar operations aboard the 63-foot-long *ASR 1357*, a Korean War-era air-sea-rescue craft. Our remote sensing fieldwork employed a Klein side scan sonar to survey the lake bottom, searching for a Champ carcass. Aboard our workboat was a remotely operated vehicle (ROV) to inspect any deepwater-sonar targets. Our vessel was motoring between Westport, New York and Basin Harbor, Vermont. At a range of about half-a-mile, toward the Vermont side of the lake, team members spotted an animate object on the lake surface. Our watercraft was going three to four knots, heading south. What we observed was black in color, no head/neck spotted, and it "churned" and "thrashed" on the lake surface, propelling itself through the water. We examined it through binoculars, too. Pat later said, "It moved slowly, would go down and then surface again a little further away. It was an independent movement from the boat wakes in the area" (Zarzynski 1988a:76).

6.2 Invergarry Castle at Scotland's Loch Oich. The waterway has had some reports of Nessie-type animals. (Credit: Joseph W. Zarzynski)

motored north on A82 toward my destination, Fort Augustus on Loch Ness. My first order of business was to find lake monster hunter Tim Dinsdale. He was not at his boat at Fort Augustus. So, I motored up A82 to Drumnadrochit to visit the new Nessie exhibit and then to go to Inverness.

Drumnadrochit is Gaelic for "Ridge of the Bridge." The settlement lies on both sides of the River Enrick, a waterway that looked more like a stream. To the west of Drumnadrochit is the village of Milton and to the south is nearby Lewiston. On my ride to Drumnadrochit, I stopped at a parking overlook that is north of Invermoriston, where I conducted shore watching for Nessie. After all, I was there to search for the Loch Ness monsters.

At Drumnadrochit, I briefly spoke to a busy Tony Harmsworth, the curator of the Loch Ness Centre & Exhibition, a museum originally called the Loch Ness Monster Exhibition. Tony moved from England to Loch Ness and in 1980, in conjunction with Ronnie Bremner, a local businessman, they co-founded the museum. Our conversation that morning was quite pleasant, though brief, as it was the beginning of a new week during the height of the summer-tourist season.

6.3 The Loch
Ness Monster
Exhibition in
1981. (Credit:
Joseph W.
Zarzynski)

After our chat, I was back in the sedan, heading the 16 miles to Inverness. When I arrived, I changed into running shorts and ran a brisk-four miles, doing my best to dodge the busy-vehicular traffic on the winding streets of the city. At the time I did not realize that this type of evasive jogging on the crowded byways of Inverness would be excellent practice for avoiding cars and lorries during my August 1984 ultrarun over the pavement of route A82.

After my jog in Inverness, I washed up at the motor lodge and found Tim Dinsdale. He had just returned from a trip to Loch Morar, another famous water-way in Scotland known for having lake monsters. We talked "shop" for awhile, then we drove over to a nearby parking lot that overlooked the loch. We got in 45 minutes of monster watching. It was a grand time sharing the watch with the "master."

Tony Harmsworth had invited both Tim and myself to dinner at the Harmsworth house in Drumnadrochit. Wendy, Tony's wife, was charming and it was relaxing, spending time socializing and talking about the monster phenomenon. Following dinner, Tim and I went to the bar at the "Drum," the nickname for the Drumnadrochit Hotel, for an after-dinner lager before I headed to my lodging for the evening.

The following day, July 14, Tim Dinsdale and I visited Jim Hogan at his Caley Cruisers facility along the Caledonian Canal, just north of Loch Ness. Jim owned a thriving boat business that rented spacious cabin cruisers to the public to ply the picturesque waters of the Great Glen. Tim knew Jim Hogan from the many times they saw one another while boating on Loch Ness, Tim scanning the waters looking for Nessie, and Jim delivering boat rentals or making on-the-water vessel repairs to his rental flotilla.

From 7:00 p.m. that evening until 2:00 o'clock in the early morning hours of the next day, July 15, I worked with Jim conducting sonar fieldwork on the loch. We were aboard his favorite vessel, the *New Atlantis*. Hogan had converted the craft for monster hunting by investing £15,000 ($19,650 in 1984, equivalent of $41,218 in 2020) into remote sensing equipment. The fiberglass-cabin cruiser, over 30 feet in length, was rigged with impressive sonar gear. This included a Simrad AR650, a Simrad SY, a Kelvin Hughes transit sonar (the transducer being pole mounted, looking out to one side of the vessel), and a Simrad Skipper 603.

Jim said he would frequently sonar scan the depths of the waterway in search of its mystery creatures. Though we were not lucky enough that evening and early morning to pick up any sonar record of a deep-diving Nessie, the boat excursion certainly was near the top of my list of the best things about my 1981 Loch Ness expedition. I have to admit, I was a junkie for fieldwork and this sure was exciting.

It was July 15, and after getting back to the hotel and grabbing a few hours sleep, I was able to get in a 5.7-mile jog along the roads of Drumnadrochit. I had finished my first marathon in May 1981, and I planned on running another one in the autumn. I could not miss many days without jogging.

Later in the day I rendezvoused with Ivor Newby, a fellow Loch Ness monster hunter, whom I first met in 1979. He was known to Nessie aficionados as "Ivor the Diver," since he did scuba diving in the waterway during research projects. He lived in the West Midlands of England. Ivor would occasionally trek north to Loch Ness to help out on Nessie searching and to see friends who regularly migrated for the "monster hunting season." Ivor knew Tim Dinsdale as well as other Academy of Applied Science researchers. As a young man, Ivor competed in long-distance car races in exotic places around the world. He even owned a couple of amphicars.*

---

*Amphicars were amphibious automobiles, also known as aqua cars. They first went into production in the early 1960s. These multi-purpose vehicles were marketed only for several years. About 14 ¼ feet long, they could operate on *terra firma* and in water, too. In water, each amphicar was powered by two propellers located in the rear of the vehicle.

6.4 Ivor Newby, a veteran-Loch Ness monster hunter, and Pat Meaney in Newby's amphi-car at Stratford-upon-Avon, England in 1982. (Credit: Joseph W. Zarzynski)

The following year, in 1982, Ivor took Pat and myself on an amphicar ride upon the River Avon at Stratford, England. Ivor was also a talented singer and he even had some minor acting roles on British television. He worked closely with the Loch Ness Phenomena Investigation Bureau during the 1960s into the 1970s.

In the evening of July 15, 1981, Ivor and I drove to Tychat Cottage on the northeast side of Drumnadrochit, overlooking Temple Pier, to have dinner with Dr. Robert and Carol Rines. Bob Rines had been an undergraduate student at the Massachusetts Institute of Technology. He later earned his law degree from Georgetown University. Rines also received a doctorate from Taiwan's National Chiao Tung University. The Boston-born Rines was an inventor, too. As an academic, Rines founded the Franklin Pierce Law Center in 1973, which became the only law school in New Hampshire (Martin 2009). He likewise directed the Academy of Applied Science, the organization that was attempting to explain the mystery of Nessie using cutting-edge underwater technology like sonar, hydrophones, and underwater cameras.

During dinner conversation, Rines reviewed the AAS's projects at the loch. He then invited Ivor and myself to help him prepare a custom-crafted underwater-camera system developed by Harold Edgerton. The camera array would be affixed to a floating platform located offshore in Urquhart Bay. We quickly agreed and Dr. Rines informed us the work would be undertaken in two days in the makeshift workshop that was adjacent to his Drumnadrochit residence.

6.5 A camera platform anchored offshore in Urquhart Bay, Loch Ness. This flotation was used to deploy a Nessie-searching camera designed by "Doc" Harold Edgerton of the Massachusetts Institute of Technology and the Academy of Applied Science. (Credit: Joseph W. Zarzynski)

The following day, July 16, 1981, upon the advice of Tim Dinsdsle, I drove 48 miles to Loch Quoich, another of the "maybe-monster" waterways in Scotland. It was informative, visiting another of the reputed monster waters of Scotland. Could it be that one or more of these other lochs might likewise be the home(s) of Nessie-type creatures?

Upon returning to my motel room at the lodge next to the Drumnadrochit Hotel, I changed into running gear and jogged 4.7 miles around the roads of Drumnadrochit. Following my workout, I showered and drove to the east side of Loch Ness and to the village of Dores. Adrian Shine of the Loch Ness & Morar Project, a group that was formed out of the old Loch Ness Phenomena Investigation Bureau, was on the loch on their motorized pontoon vessel. One of their tasks that summer was to assist Heriot-Watt University of Edinburgh in a remotely operated vehicle (ROV) survey of the sunken Loch Ness Wellington bomber that was discovered in 1976.

A ROV is a tethered underwater vehicle, generally equipped with a camera system that is operated by a pilot who is on the surface. In 1981, this was state-of-the-art technology for underwater exploration and survey. Their underwater-videography survey, using the ROV Sea Pup, imaged the sunken World War II aircraft. The operation was directed by Robin Holmes, an instructor in the Department of Electrical and Electronic Engineering at Heriot-Watt University. Holmes had secured funding for the project from the Royal Air Force Museum in Hendon, United Kingdom and from *FlyPast*, a vintage-aircraft magazine based in Britain. On what Robin Holmes would later call a "shoestring budget," his team spent July 13-24, 1981, deploying a ROV to visually document the structural condition of the rare-military aircraft lying on the loch bottom at a depth of over 200 feet (Holmes 1991:9-10). I was unable to talk with Adrian Shine on July 16, 1981, but I was later informed that Shine graciously offered the Loch Ness & Morar Project's pontoon boat and his group's services at no cost to Robin Holmes and Heriot-Watt University (Holmes 1982:58-59). Who says lake monster hunters didn't make real contributions for societal advancements?

After breakfast the next day, July 17, 1981, I drove to Bob Rines' cottage to assist in prepping his group's underwater camera for stationing in the depths of the

6.6 Sea Pup, a remotely operated vehicle, aboard a research boat at Loch Ness in 1981. (Credit: Joseph W. Zarzynski)

loch. Ivor Newby soon arrived and the three of us labored on the customized un-derwater camera stored in a makeshift lab in Rines' garage. After a few hours of tinkering, the Edgerton camera was ready to be lowered into Urquhart Bay. "Doc" Harold Edgerton, a professor at MIT, was instrumental in the development of strobe photography and early underwater cameras. Thus, Edgerton was known as "Papa Flash" by Jacque Cousteau's research divers aboard the vessel *Calypso*.

The three of us then boarded the vessel *Hunter*, docked at Temple Pier, and we motored out to a floating platform in Urquhart Bay. For two hours we readied the photographic gadgetry. The camera apparatus, with its strobe lighting, was then suspended from the raft anchored in the bay. I was told the camera would periodi-cally snap underwater images in the hope that an inquisitive Nessie creature might swim by and be photographed.

It was almost dinnertime when Ivor and I returned to our respective lodgings, he in the "Drum" (aka Drumnadrochit Hotel) and me next door in the hotel's other guest residence, the motor lodge. After dinner we had a couple of beers at the bar in the "Drum," discussed our fieldwork with the AAS from earlier in the day, and then we retired for the evening. It had been quite a productive day at the loch.

6.7 A close-up photograph of the Academy of Applied Science's floating-camera plat-form anchored in Urquhart Bay, Loch Ness in 1981. (Credit: Joseph W. Zarzynski)

I got up at 6:00 a.m. on July 18, dressed into my running gear and drove the 25 minutes to Inverness for an early-morning jog. I wanted to see the capital city of the Scottish Highlands one last time before I would depart Loch Ness. I was able to run 3 ½ miles before the Inverness-commuter traffic got too heavy.

After my jog and a shower back at the lodge, Ivor and I ate breakfast together before we again visited with Bob Rines. However, the AAS chief was too busy for a long chat. Since my time was limited, I said my goodbyes to Ivor and Rines, departed Tychat Cottage, and I checked out of the lodge. I drove off to Lochs Arkaig, Shiel, and Morar, three other of the so-called monster haunts of Scotland. These were very short visits, but I did meet Jim Penny, the water bailiff of Loch Morar. Jim was not only a superb waterway caretaker, he also was the one person you wanted to talk with to find out if there were any recent sightings of "Morag," the purported Nessie-like animals of Loch Morar.

It was then off to Oban, a seaport on the west coast of Scotland. There, I secured a room at an inn for the evening. The next morning, I was able to do some sightseeing in Oban and I also found time to get in a 4.2-mile jog around the historic resort town. After cleaning up, I checked out of the hostelry, drove down near Prestwick Airport, and stayed the evening of July 19 at a hotel in Ayr. I went to bed early that night since July 20, 1981 was my flight back to JFK Airport. The long-jet ride from Scotland to New York City was followed by a 3 ½-hour vehicle drive back to my tiny apartment in Saratoga County. It had been a rewarding expedition to Loch Ness and I had many tales to relate to my girlfriend Pat.

Chapter 7

Other Scottish Loch Monsters

In the last chapter, I briefly mentioned some of my expeditions to Loch Ness including trips taken to other Scottish waterways that had reports of unidentified animals having been observed by Highlanders. Ivan T. Sanderson (1911–1973) was a pioneering cryptozoologist. The Scotland-born biologist and naturalized American was one of the first researchers to notice that in the northern- and southern-hemispheric lands, between 40- and 60-degrees latitude, there were several lakes that had reports of unknown aquatic animals (Coleman and Huyghe 2003:267-270).

Based upon Sanderson's observation and a suggestion to me by Tim Dinsdale in the summer 1982, Pat Meaney and I visited several of Scotland's other alleged lake-monster waterways. The year 1982 was also a high-water mark for monster hunting in general.

Only a few months earlier, a formal association was established known as the International Society of Cryptozoology (Greenwell 1982:1). The International Society of Cryptozoology (ISC) was founded in January 1982 at a meeting of its board of directors at the Smithsonian Institution in Washington, DC. That convention of scientists was hosted by Dr. George Zug and the Smithsonian's Department of Vertebrate Zoology. Dr. Bernard Heuvelmans became the society's president. The ISC was established to: 1) publish articles and reports on cryptozoological topics, 2) exchange information about cryptozoology, 3) encourage scientists to investigate hidden animals without apprehension that their work would be controversial, 4) create an organization in which the public and media could seek advice on the topic, 5) help legitimize the subject of cryptozoology, and 6) serve as

Map of Scottish Monster Lochs

North Sea

Atlantic Ocean

■ L. Assynt

L. Quoich   ■ L. Ness
■ ■ L. Oich
■ L. Lochy
L. Morar ■ ■ L. Arkaig
■ L. Shiel

SCOTLAND

ENGLAND

7.1 Map showing the locations of several lochs in Scotland, each with a history of having reports of water monsters. (Credit: Joseph W. Zarzynski)

a historical archive for scholars to store their cryptozoological files and reports (Greenwell 1982:1).

After nearly a quarter-century, the ISC ceased operation. Then in 2016, a similar organization, the International Cryptozoology Society (ICS), was founded in St. Augustine, Florida during a meeting of cryptozoologists organized by Portland, Maine's Loren Coleman (Coleman 2016).

Therefore, it seemed appropriate that Pat and I journey to Scotland in the summer of 1982 to examine some of the other waterways reported to have Nessie-like animals. Pat was a native of Stony Brook in Suffolk County, New York, on the north shore of Long Island. She grew up cultivating an innate curiosity to investigate "things." As a youngster, Pat and her younger brother John spent hours walking the Long Island beaches around their hometown looking for nature's anomalies. She was captivated by the variety of sea life literally just down the street from her family residence. Likewise, Pat wondered if Captain Kidd had buried pirate treasure along her neighborhood's sandy shoreline as was believed by many Long Islanders. Thus, it is not surprising that Pat's youthful zeal to explore intriguing topics was later manifested by her profession, a librarian who loved reading mystery books. Pat was the junior high school librarian in the same building where I taught.

Our task during the summer of 1982 was to undertake a more in-depth reconnaissance of several little-recognized cryptozoological waterways. Each of these waters had a reputation, let's call it a legend, of being the habitat of hidden animals.

7.2 Lake monster hunter Pat Meaney at one of the Scottish lochs in 1982. That summer, Meaney and the author visited several waterways in Scotland that had reports of lake monsters.(Credit: Joseph W. Zarzynski)

In the inaugural edition of *Cryptozoology,* a scientific journal published in 1982 by the International Society of Cryptozoology, Pat and I authored a formal report on our 1982 fieldwork in Scotland. The article was entitled "Investigation at Loch Ness and Seven Other Freshwater Scottish Lakes" (Zarzynski and Meaney 1982:78-82).

It would be misleading if I professed that all of these Scottish lochs were the habitats of cryptids. However, I nonetheless recognized that these waters deserved a preliminary examination. The phenomenon of "water horses," a common moniker for the unknown animals in Loch Ness and other Scottish waterways, is part of the cultural psyche of many Highlanders. In addition, our 1982 fieldwork was in some respects an investigation that primed me for my Loch Ness ultrarun two years afterwards.

From August 6-19, 1982, Pat and I traveled around the United Kingdom undertaking surface surveillance at Loch Ness and seven other Scottish waterways using the basic equipment of cryptozoologists at the time—binoculars and cameras. This expedition was sparked a year earlier, during my 1981 visit to Loch Ness. As previously mentioned, on July 13, 1981, Tim Dinsdale, Tony Harmsworth, Tony's wife Wendy, and I were having dinner at the Harmsworth residence in Drumnadrochit.

During dinnertime banter, Tim suggested that I should tour some of the other reputed-monster waters of Scotland. The veteran Nessie-seeker recommended cursory visits to "Loch Morar, Loch Lochy, Loch Arkaig, Loch Shiel, and Loch Quoich" to find out more about these lesser-known "monster haunts" (Zarzynski 1981).

Therefore, embracing Dinsdale's recommendation, in August 1982, Pat and I conducted reconnaissance surveys to these and other waterways. We hoped our expedition might establish lines of communication between American and European (specifically Scottish) cryptozoologists and also be a preparatory endeavor for a possible return to one or more of these lochs for in-depth investigative fieldwork (Zarzynski and Meaney 1982:78).

Our 1982 trip was a grand adventure, our own Professor Challenger* expedition into Scotland's water creatures. It was a faraway inquiry that both Pat and I thoroughly enjoyed, and the trip also bonded our relationship that led to us getting married in 1985.

Here is a brief synopsis of those so-called monster lochs that we visited in 1982. In 1984, we would return to a couple of these waterways for more reconnaissance work. Our field operation in 1982 gathered background information about each loch and whenever possible we collected reported sightings of their respective mystery denizens. It was quite the cryptotourism excursion.

Loch Ness is by water volume the largest freshwater lake in all of the British Isles and is most known for possibly being the "home of Nessie." It is over 22 ½ miles in length, but some researchers list it at 24 miles. The loch is about two miles wide.

Loch Morar is the deepest lake in Scotland at 1,017 feet. It is 11 miles long. Its cryptids are known as Morag. The tag is also a popular name for girls in Britain and rather appropriately, Morag means "great."

---

*Professor George Edward Challenger was a fictional-British explorer created by Arthur Conan Doyle, who likewise conceived the literary Sherlock Holmes. Doyle's novel *The Lost World*, published in 1912, included the travels of Professor Challenger who discovered a world of dinosaurs and other exotic creatures. Professor Challenger was based upon the real-life exploits of Percy Fawcett, who had explored the jungles of South America looking for a legendary lost city.

7.3 Loch Morar, Scotland's deepest lake, is the reputed home of a legendary monster nicknamed Morag. (Credit: Joseph W. Zarzynski)

Loch Assynt is over six miles in length and has a maximum depth of 282 feet.

Loch Arkaig is 12 miles long with a mean width of a half mile. Its maximum depth is 359 feet. Its hidden animals are known as "Archie."

Loch Shiel is 17 miles in length with a maximum depth of 420 feet. Its enigmatic creatures are called "Seileag."

Loch Quoich is nine miles long with a width of 1 ½ miles. Its maximum depth is 281 feet. It is, however, a reservoir and its size was increased from the construction of a hydroelectric dam. That not only expanded the waterway's length from seven to nine miles, it also broadened the loch's width and increased its depth. Therefore, it seems unlikely that this loch and reservoir is the habitat of Nessie-looking animals. Nonetheless, there have been some sightings of water monsters at Loch Quoich.

Loch Oich is the smallest of the so-called monster lochs of the Scottish Highlands being only four miles in length. The loch's mean breadth is one-fifth mile. Its maximum depth is 133 feet. One of the loch's most notable sightings of "Wee Oichy," the name for its denizens of the deep, occurred in 1936, just three years after the Mackays' sighting at Loch Ness. Mr. A. J. Robertson was boating near the southwest end of Loch Oich. He reported observing a "black snakelike body" with a "vaguely dog-like" head protruding three feet in front of its humps (Costello 1974:143-144).

Loch Lochy is 10 miles long with an average width of three-fifths of a mile and the maximum depth is 531 feet (Zarzynski and Meaney 1982:78-82). Its enigmatic critters are called "Lizzie."

Peter Costello, an Irish cryptozoologist and author, summed it up quite neatly when he wrote in his 1974 book *In Search of Lake Monsters*: "The Loch Ness animals are not unique, either in Scotland or the world. The more attention given to reports from other places, the sooner the real existence of these animals will be completely established" (Zarzynski and Meaney 1982:82).

What might these sightings be, if they are not unidentified animals, that is, aquatic monsters? In my 1984 book, *Champ—Beyond the Legend*, I addressed this for the monster sightings at Lake Champlain. The same candidates might also apply to Loch Ness and the other reported monsters said to inhabit these Scottish waterways. The following is a list of what eyewitnesses have seen and possibly misconstrued as a lake monster:

1. Unique waves
2. A floating log or logs
3. Large fish or a school of fish
4. Atmospheric refraction, where a surface temperature inversion creates optical trickery, producing a phenomenon such as a stick poking out of the water that appears to be the long neck of a plesiosaur-type creature
5. Swimming animal(s) (dog, waterfowl, otter, deer, etc.)

7.4 A Champ-looking log lying along the shore at Lake Champlain in the early 1980s. (Credit: Joseph W. Zarzynski)

The sightings of Nessie, Champ, and other water monsters might be one of these:

1. Amphibian (frog, toad, or salamander)
2. Supersized fish or eel
3. Invertebrate (squid, worm, sea slugs, etc.)
4. Reptile, such as being from the plesiosaur family
5. Mammal (whale, seal, sea lion, or dolphin) (Zarzynski 1984:97–98)

Of course there is always the possibility of hoaxes or the "old standby," lake monster sightings are due to an eyewitness being influenced by alcohol or drugs. Furthermore, you could add to this list a paranormal aspect. Tony Healy and others think the sightings of water monsters at Loch Ness and elsewhere could be related to the paranormal, creations outside the range of scientific explanation and reasoning.

Of the eight waterways mentioned in this chapter, only two, Loch Ness and Loch Morar, have had substantial cryptozoological investigation. Besides Ness, only Morar and Shiel have had more than a handful of recorded lake monster sightings. In early 1983, J. Richard Greenwell (Secretary-Treasurer and a co-founder of the International Society of Cryptozoology), Richard Smith (Wind & Whalebone Media Productions), and I (Lake Champlain Phenomena

Investigation) planned to travel to Scotland in the summer of 1983 to study two of these lesser-investigated waterways.

Our intention was to conduct three weeks of cryptozoological fieldwork with our time and efforts split between Lochs Morar and Shiel. We felt that adequate field studies had already been attempted at Loch Ness over the 50 years since the world-famous sighting by the Mackays in 1933. We believed it was time to examine these other waterways in greater detail. We planned to conduct surface- and underwater-video and still-camera surveillance, scuba reconnaissance, and other fieldwork. The expedition would have been executed under the sponsorship of the Tucson, Arizona-based International Society of Cryptozoology. Richard Smith hoped that this operation would be noteworthy enough to attract television-production companies to sponsor a documentary of the proposed project (Zarzynski 1983). Because of the absence of outside funding, however, our 1983 fieldwork to Loch Morar and Loch Shiel never happened.

Chapter 8

1984—A Most Interesting Year

By the summer of 1984, I had been involved in cryptozoological studies at Loch Ness and Lake Champlain for nearly a decade. Further, my book *Champ—Beyond the Legend*, published by Bannister Publications (Port Henry, New York), was released in July 1984. The book was an outgrowth of a 1983 article I had written for *Adirondack Bits 'n Pieces*, a regional magazine published by Bannister Publications. My story in the inaugural issue of the magazine was entitled "'Champ'—A Zoological Jigsaw Puzzle." The feature was well received by the public and the publisher, J. Robert DuBois (Zarzynski 1983:16-21).

I first met Bob DuBois in November 1980, when I brought *New York Times* reporter William E. Geist up to Port Henry, New York, where DuBois lived and worked. Geist was researching an article for the *New York Times* on the Lake Champlain monsters and their impact on tourism for the lakeside hamlet of Port Henry, known as the "Home of Champ." Years later Bill Geist would be hired as a correspondent by CBS. Geist would eventually win two Emmy awards in television and he also received a star on the Hollywood Walk of Fame (Encyclopedia.com 2020). In 1980, Bob DuBois operated a hotel in Port Henry, an area that once was known for iron-ore mining. DuBois was one of many residents of Port Henry interviewed by Geist. Bob DuBois remembered me from that 1980 trip and in 1983, he asked if I would write an article on Champ for the first issue of *Adirondack Bits 'n Pieces* magazine. Following the success of that initial issue, Bob invited me to author a book on the topic of Champ, that Bannister Publications would publish.

I vividly recall driving up to Port Henry on July 19, 1984 to pick up my publisher for our two-hour drive to Capital City Press, located near Montpelier, Vermont.

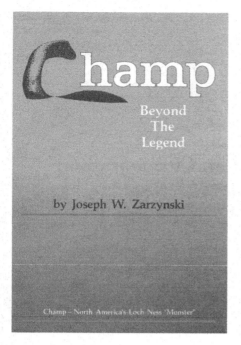

8.1 The front cover of Joseph W. Zarzynski's 1984 book Champ—Beyond the Legend. *(Credit: Joseph W. Zarzynski/Bannister Publications)*

Champ—Beyond the Legend had just been printed by the press and Bob DuBois, the book's publisher, wanted to see the first copies of the publication. I had a cargo van and Bob likewise desired to pick up part of the printing run of the publication, rather than wait several days for a shipment of books to reach him in Port Henry. As we crossed the Crown Point Bridge from New York over Lake Champlain into Vermont, Bob seemed edgy and nervous. I inquired how he was feeling. My publisher replied that he had a sleepless night due to a "nightmare," as he described it. Bob confessed that his anxiety was about the book and its front cover. He dreamed it had an image of a candy wrapper printed over the title. Bob said his nighttime vision was of a discarded candy wrapper that had fallen into the printing press at the very moment the book's cover was being printed. I guess for a publishing-house proprietor, that would have been an appalling thought. Suddenly, I, too, began wondering about the abhorrent dream.

The first thing we did when the press's manager showed us the cases of books stacked up in their warehouse, was to check one of the publications. When we inspected the newly printed book, there was no candy wrapper stamped over the front cover. I think I never saw Bob happier. He had a broad grin of relief on his face, even though he was about to write a check for a substantial sum of money to pay for the printing. The well-respected press did a terrific job, too, with the production of the book.

So, 1984 was an exciting year in my life. I got engaged in April, my first book was published in July, and my ultrarun along the full length of Loch Ness was scheduled for August.

Chapter 9

1984—Prepping for an Ultramarathon and Going to Loch Ness

By 1984, I had been running marathon races for three years. In my mind it was a natural progression to combine two principal interests, searching for the Loch Ness monsters and distance running. Therefore, early in 1984, I began formulating plans for an ultramarathon of Loch Ness. I had looked for lake monsters using binoculars and cameras, sonar gear deployed from boats, sonar affixed to a tripod and placed underwater, and scuba dives searching for a carcass of a Champ animal. Running along Loch Ness would be just another way for me to get figuratively closer to these unknown animals. I thought I might even see a Nessie during my ultrarun. Moreover, "runners run" and cryptozoologists "look for hidden animals." Therefore, I just combined the two!

9.1 Joseph W. Zarzynski at Lake Champlain with scuba gear standing next to his work van in 1982. (Credit: Pat Meaney)

I was in great physical condition leading up to that solo ultramarathon. My exercise training in the months before the long-distance run consisted of moderate jogging combined with plenty of cross training, the latter doing various types of exercises to work all parts of the body. My cross training consisted mostly of weight lifting, basketball, scuba diving, and even pool swimming. I believed the weight training was integral as it would strengthen my legs and the upper-body muscles needed to be an effective ultrarunning athlete.

However, I was concerned about my meals in Scotland. I have been a vegetarian since I was two years old. All my life I have on numerous occasions struggled trying to find meals when traveling from home, especially in a meat-loving country like Scotland.

By 1984, I had seven marathons "under my belt" with the last one completed 10 months before my Loch Ness ultrarun. Sometimes, training for distance-running events was problematic for me, mostly a case of finding opportunities to exercise and also to get enough rest to recover from workouts. I was a school teacher and personal time during the September-to-June school year was frequently scarce. My work day as a teacher included preparation of lesson plans, instructional time, meetings with colleagues and parents, and of course, grading homework, quizzes, and tests. Like most teachers in my school district, the typical workday was about 10 to 11 hours each Monday through Friday, with several hours over each weekend devoted to preparing lessons and grading. Managing one's time to fit in exercise workouts was a challenge, but this was also the norm for anyone who was a marathon or ultramarathon competitor. Getting enough sleep also seemed to be an issue. In addition, I lived in the foothills of the Adirondack Mountains in upstate New York, a place where it snowed a lot from November into March. Therefore, jogging outside in the winter months was iffy at times and I disliked running indoors on a treadmill.

My training for the several months leading up to my August 22, 1984 Loch Ness run, including the full month of August, consisted of:

January 1984: (20 runs, long run that month—10.4 miles, 111.1-miles total, 9 basketball sessions, and 12 weight training workouts)
February 1984: (19 runs, long run that month—15 miles, 100-miles total, 5 basketball sessions, 9 weight training workouts, and 1 pool swim)

March 1984: (22 runs, long run that month—18.6 miles, 112.3-miles total, 6 basketball sessions, 11 weight training workouts, and 1 pool swim)

April 1984: (23 runs, long run that month—10.3 miles, 118.4-miles total, 6 basketball sessions, and 9 weight training workouts)

May 1984: (27 runs, long run that month—two runs of both 10.1 miles, 124.9-miles total, and 11 weight training workouts)

June 1984: (25 runs, long run that month—8 miles, 107.9-miles total, 9 Nautilus weight training workouts, and 1 scuba dive)

July 1984: (26 runs, long run that month—14.5 miles, 153.8-miles total, 12 weight training workouts, and 4 scuba dives)

August 1984: (18 runs, long run that month—28.5 miles, 111.3-miles total, 4 weight training workouts)

I was in pretty good shape for my upcoming ultrarun, even though my training had not been as great as for some of my marathon races. My height was nearly 6 feet, 6 inches and I weighed 175 pounds, a healthy marathoner and ultramarathoner.

One of the things that concerned me with my cryptozoology fieldwork in the first half of 1984 was I was not making as many scuba dives as normal. Scuba diving was not just a sport, it was a craft as well. To excel in it you needed to make many dives to gain as much experience as possible. Fortunately, late summer and autumn in the northeast USA were splendid times of the year for scuba diving. I figured I would have numerous opportunities after my ultrarun to conduct cryptozoology-scuba forays at Lake Champlain.

In the seven weeks before my Loch Ness run, I completed seven training runs over 10 miles. In early August, 17 days before my Loch Ness ultramarathon, I completed a 15.5-mile jog followed up several days later with a run of 10.3 miles. I would have preferred to have gotten in a couple of 20- to 22-mile jogs, which I generally did as part of my training before marathon races. However, due to time constraints I was unable to accomplish that. Nevertheless, I was in excellent condition for distance running.

My training pace during these 1984 jogs was generally between 7 minutes, 30 seconds and 8 minutes per mile. My main goals in the weeks leading up to my Loch Ness ultrarun were to keep physically active, run whenever possible, and do lots of

9.2 Pat Meaney (center) and the author (left) scuba diving at Button Bay, Lake Champlain in 1981. The pair tested newly acquired sonar for cryptozoologist Dr. Roy P. Mackal (right). The gear was used later that year in Central Africa in a quest to find the reported water beastie called mokele-mbembe. The unknown animal has been described as looking "half elephant, half dragon." (Credit: J. Richard Greenwell)

cross training. What's more, I turned 34 years old in July 1984, so I was just entering my prime as a marathoner and ultramarathoner athlete. Our trip to Scotland was to begin on August 16, 1984. As Pat and I packed for our overseas journey, I felt I was ready for my ultrarun at Loch Ness.

Thursday—August 16, 1984

In the early morning of August 16, 1984, Linda and Art Kranick, a couple of teaching colleagues who would become two of the most accomplished high school cross country coaches in the country, drove Pat and me from our residence to a suburb of Albany, New York. There we boarded a shuttle van for transport to JFK Airport in Queens, New York. However, the flight to Scotland was soon plagued by a couple of unexpected problems.

First, about halfway into the passenger-van ride to JFK Airport, I realized I left my prescription eyeglasses at home which I needed for driving a vehicle. Since we would have a rental car in Scotland, I absolutely required corrective eyewear. After that discovery, Pat and I simply did not have enough time to return to our apartment in Saratoga County, retrieve my eyeglasses, and then to get to JFK Airport in time to make our flight. I decided we would have to improvise when we arrived in Scotland. Somehow I would find a suitable pair of eyewear. I had not come this far in my preparation for running an ultramarathon to be sidelined by a pair of missing spectacles. However, as our land transport entered the property of JFK Airport, my stomach turned from the uncertainly of how to remedy this complication.

After checking in, we were notified that our passenger jet to the United Kingdom would be delayed for two hours due to a thunderstorm near the airport and also because of heavy air traffic. Finally, following that unexpected delay, our flight departed New York. After several hours in the air we landed at the Manchester Airport in central England. Our final leg of the flight was north to Prestwick, an airport outside Glasgow, Scotland.

I had flown into Prestwick a few times and enjoyed this less-used airfield located in a non-urban surrounding. The airport originally was constructed in the mid-1930s and after the end of World War II in 1945, it began transatlantic service to-and-from New York. Moreover, the countryside environment and lower volume of road traffic around the airstrip provided me with a few minutes to adjust to driving on the opposite side of the road. In the United Kingdom, vehicular traffic was on the left side of the roadway. I was certainly familiar with the British system of driving from my previous trips to Scotland. Thus, operating a car on the left side of the road was not a major problem for me, once I got a few minutes of experience with those driving conditions.

Friday—August 17, 1984

Upon landing in Manchester, our jet's passengers were informed there would be a one-hour delay before departing for Prestwick. That one hour soon turned into an eight-hour delay due to what the airline termed "technical problems." It was now Friday, August 17, and it was nearly dark when our aircraft finally landed at the

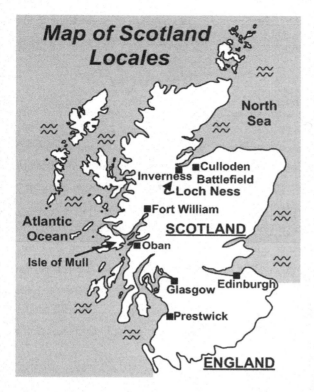

9.3 Map of Scotland Locales. (Credit: Joseph W. Zarzynski)

Prestwick Airport, located about 35 miles from Glasgow. After touching down on the aviation field, we quickly collected our baggage, cleared customs, and secured our car rental. I thought, maybe things were finally beginning to change.

Pat and I rented a comfortable sedan with plenty of leg room for a tall person like myself. Extremely tired from the long flight and delays, we exited the car-rental lot that was adjacent to the airdrome. We finally began our ground trek north toward Loch Ness. It was early nighttime when we pulled out of the airfield grounds. I was behind the wheel, but I soon had to turn the driving over to Pat due to my absence of corrective eyeglasses. Simply stated, even if I squinted, my long-distance eyesight was not adequate.

It was now dark as Pat drove. She was having some problems staying in the left-hand lane of the roadway due to a lack of familiarity with that driving style, combined with tiredness from the monumental flight delays. Discretion finally won out. We stopped at a hotel and got a room for the night. We agreed, the next day,

Saturday, August 18, would be a frantic hunt for prescription eyewear for me, the "forgetful one." Little did we realize then, that quest would nearly be as challenging as pursuing Nessie.

Saturday—August 18, 1984

We woke up early on August 18, our bodies were still immersed on east-coast USA time. Jet lag had clearly settled into both of us. After breakfast we loaded the trunk of our sedan with our luggage. Pat again was at the wheel of the automobile. This time she was doing much better navigating the unfamiliar road network of Scotland.

First we journeyed to Largs and then to Greenock, two coastal towns in western Scotland. We located a couple of optician offices, seeking distance eyeglasses for me, but both offices were closed for the weekend. Therefore, we decided to drive to nearby Glasgow, the most populous city in Scotland with over a million residents, to find corrective eyewear for me. I felt terrible that I was the cause for our delay getting to Loch Ness. Fortunately, Pat did not seem to mind too much.

In Glasgow, we parked our car and approached a police officer stationed on a street corner. We hoped he might have a suggestion on how best to overcome our major obstacle. In a Glasgow dialect, the constable responded, "You must go to the Barras." We quickly discovered that was a popular flea market in Glasgow. Mercifully, the Barras was nearby.

True to our lack of luck, we got lost navigating our way, but an elderly Scotsman walked us the several blocks to the Barras. After thanking him, we discovered that the Barras was one of the most famous outdoor markets in the world. It had stalls and tables that seemed to sell just about everything from clothing, to tools, to replica artwork, and yes, even previously owned eyeglasses. Finally, our luck seemed to have changed. We were later told by some Scots that if you were in Glasgow and ever had something stolen, you might be able to find it at the Barras, where you could buy it back.

"Eureka!" After a few minutes of rummaging through the numerous aisles, we found a stall that sold only eyeglasses. They had an impressive inventory of used eyewear of every kind, color, shape, and prescription. Next came the formidable task of trying on a few-hundred eyeglasses hoping that at least one would match my prescription.

After nearly giving up, I eventually discovered a pair of eyeglasses that matched perfectly, but it came with a bit of a "price." It was a pair of woman's sunglasses, pointed at each end of the frame, in what was an exaggerated "cat-eye style." These were popular in the 1950s and 1960s. Yet, it was 1984, and this design was not now in fashion. Moreover, the plastic-eyewear frames were a flashy-light green, yet the prescription was an absolute match. So, after some bargaining with the proprietor, I purchased them at the clearance price of one pound ($1.32 in the 1984 exchange rate). The entrepreneur was pleased with her sale as I hid my tremendous joy, since I would have paid 50 times that amount. I was now ready to navigate the urban and country roads of Scotland and Pat seemed more ecstatic then me. My fiancée could now enjoy being the passenger to view the pastoral countryside of Scotland, a land she had visited only once before, in 1982.

It was now 1:30 p.m., but we were back in our vehicle and soon were crossing Erskine Bridge in Glasgow, motoring over the River Clyde. We were leaving central Scotland, heading north toward Loch Lomond and then on to the Scottish Highlands. We arrived at Fort William around dinnertime, ate, and drove on, finally arriving at our destination, the village of Drumnadrochit on the shores of Loch Ness. It was then 8:00 p.m.

We checked into the Benleva Hotel, a structure that was reported to be over 300 years old, as well as once being a clergy's residence (The Benleva 2019). We both knew that after a good night's sleep we'd be ready to begin tackling the three-major goals of the trip: 1) to search for Nessie and the so-called aquatic beasties of other Scottish lochs, 2) prepare the logistical work to get ready for my Loch Ness ultramarathon, and 3) the solo run itself.

Sunday—August 19, 1984

After breakfast at the Benleva Hotel in Drumnadrochit, we headed over to the monster exhibit on the grounds of the Drumnacrochit Hotel in the center of the village, less than a mile from the Benleva Hotel. Our inn was a couple-hundred yards off route A82. The car drive from the hotel to the center of town was striking as we motored along the tree-lined roadway to route A82. If those woodlands could speak, I think they would have chuckled at my most-unusual spectacles, a fashion statement that rivaled the eclectic wardrobes of Elton John.

In 1984, the Drumnadrochit Hotel complex was into a rebranding, remodeling, and expansion mode. The stone-structure inn* had a few years earlier been supplemented by a motor lodge and a monster museum. The hotel, motor inn, and exhibition hall were owned and managed by Ronnie Bremner and his family.

Ronnie Bremner was an astute businessman. I was told he had been a fine tennis player in his youth and that over his lifetime he enjoyed other sports, too—golf, sailing, rugby, and billiard games. During one of my earlier visits to Loch Ness, all the local hotels in the village were full. Ronnie came to the rescue. His wife and he let me stay in a spare room in their family residence on the grounds of the hotel complex. During that visit, Ronnie took me to play a couple of games of a British-cue sport. As one might say in the United States, "He cleaned my clock," thoroughly trouncing me. So, I can personally attest to his sporting talents.

Cryptozoologists frequently patronized Bremner's lodgings and museum complex. For years, the Drumnadrochit Hotel's bar was a prominent gathering place, too, for those seeking to find other monster hunters to quaff a beverage and discuss the latest news about the "water beasties."

The monster museum on the property of the Drumnadrochit Hotel opened in 1980. Tony Harmsworth, one of the principals of the Loch Ness Centre & Exhibition (originally called the Loch Ness Monster Exhibition), said it took about 10 months to set it up. In July 1981, Tony informed me that the facility had "about 50,000 people come through…the first year" (Zarzynski 1981). The original exhibit mainly consisted of informative displays about Nessie and the monster-hunting colony who had done fieldwork at the loch. There was also a section in the gallery hall providing a primer about the waterway's limnology, that is, its biological and chemical composition. The museum's focus was initially pro-monster hunting and slanted that the loch might indeed be the home of hidden animals.

Tony Harmsworth professed during a July 13, 1981 conversation that he thought "what's in Loch Ness must be either—a pleasiosaur, a fish or a type of eel"

---

*The Drumnadrochit Hotel had a fire in January 1984 and the stone building was then renovated.

(Zarzynski 1981). Harmsworth would go on to research and write several informative books and booklets about Loch Ness.

Adrian Shine was one of the local monster hunters. Tall, lean, and bearded, he looked like the very stereotype of a homegrown Nessie searcher. Around 1973, he became one of the principals of the Loch Ness & Morar Project. The "Project," as it was sometimes called, conducted scientific fieldwork at Loch Ness and also at another Scottish lake, Loch Morar, located near the west coast of Scotland. Loch Morar was the home of the legendary creature known as Morag (Campbell and Solomon 1972).

I had met Adrian Shine on a few occasions during my early field trips to Loch Ness. The native of Surrey, England started out as a Nessie and Morag searcher, but later began referring to himself as a scientist or naturalist. In 1984, I don't think Adrian was much interested that I was a cryptozoologist, since he seemed to have a healthy dose of skepticism. However, Adrian once commented to me that he was impressed that I was a scuba diver.

In 2006, years after my ultrarun at Loch Ness in 1984, Adrian Shine would star in a Toyota Tacoma television commercial that aired periodically in the

9.4 Nessie hunters Adrian Shine (with beard) and author (at right) aboard a work vessel at Fort Augustus, Scotland. (Credit: Pat Meaney)

United States. The "small screen" commercial employed a likeness of Loch Ness as its setting. With his long beard, chiseled facial features, and charismatic British accent, Adrian was splendid in the role of a monster stalker. Pat and I saw the television commercial numerous times on American television, and we hoped that Shine was making a lot of money for his wonderful acting performance.

During our stop at the Loch Ness Centre & Exhibition on August 19, 1984, Pat and I toured the museum and we snapped numerous photographs of its excellent displays. We then visited Tony Harmsworth in his office. I presented Tony with an autographed copy of my new book, *Champ—Beyond the Legend*. The three of us then ate lunch in the hotel's dining room, chatting about the museum and Nessie, between mouthfuls of our meal.

Following lunch, Pat and I drove down route A82 to Fort Augustus at the south end of Loch Ness. I wanted to motor along route A82 as many times as possible, so that both of us became acclimated to the twists and turns in the roadway and we knew the locations of pull offs for parking.

Back at the inn, I donned running gear and did a 5.4-mile jog around the narrow roads of Drumnadrochit. Those streets had plenty of sharp turns and hardly any shoulders, so running was somewhat dangerous due to vehicular traffic that would suddenly appear from around a roadway curve. Nevertheless, it was a suitable tune up for my upcoming ultrarun.

For dinner, Pat and I went to the Inchnacardoch Hotel, located just north of Fort Augustus. It was one of the best meals we had during our 1984 trip to Loch Ness. Our dining table was near a window overlooking the south end of the breathtaking loch. The excellent food combined with an incredible landscape was well worth it. After dinner we returned to our lodging in Drumnadrochit.

Monday—August 20, 1984:

Following breakfast, Pat and I drove south on A82 to determine where my ultramarathon would actually begin. We decided it would start in Fort Augustus and end beyond the north boundary of Loch Ness, past Lochend. That way I would be running north, heading into approaching traffic. The name Lochend was certainly fitting for the finish line for an ultramarathon along the expanse of Loch Ness. I was then really getting excited about the run, in fewer than two days.

After deciding on the spot for the "starting line" for my ultrarun, we departed Fort Augustus and headed north on route A82, past Lochend, to my self-designated "finish line." We then drove further down the road and visited with Jim Hogan, the successful proprietor of Caley Cruisers on the Canal Road in Inverness.

For nearly a decade-and-a-half, Jim and his wife had owned a boat-rental enterprise based near the north end of Loch Ness. The Hogan family business had a few-dozen liveaboards in the Caley Cruisers fleet of cozy vacation boats. We chatted and I presented Jim with a copy of my recently published Lake Champlain-monster book. Not only was Jim Hogan an established and hardworking entrepreneur, as I had discovered during my 1981 trip to Loch Ness, he also was very much intrigued by the Nessie phenomenon.

Following our chat with Jim Hogan, Pat and I had a lot of daylight left. Therefore, we did some cultural sightseeing away from Loch Ness. First we visited Fort George, an impressive stone fortification that was about 11 miles northeast of Inverness. The British fortress was completed in 1769 (Johnson 2020).

Following our tour of Fort George we motored 10 miles to the site of the Battle of Culloden, a venerated tourist destination in Scotland. For many Scots and others as well, Culloden is not just an historic site, it is akin to standing on sacred ground. The Battle of Culloden (April 16, 1746) was where the Duke of Cumberland and his British-government army crushed the Jacobite uprising of Bonnie Prince Charlie (aka Charles Edward Stuart). About 1,250 Jacobite followers, an army consisting of Scots, some French, and other supporters, and 50 English soldiers died in ferocious combat (National Trust for Scotland 2020). Many of the Scots that survived the fighting and others who supported the cause of the Jacobites, fled their homeland following this defeat. Those Scottish refugees migrated to various places around the European continent and also to North America.

Pat and I departed the historic site, saddened by what happened at Culloden in 1746. We drove toward General Wade's Military Road, a thoroughfare that was a single lane and in sections it opened into a double lane. The roadway somewhat paralleled the eastern side of the loch, as we headed south, toward Fort Augustus. It had been an uncharacteristically hot day for August as temperatures hit into the

9.5 General Wade's Military Road on the east side of Loch Ness. (Credit: Joseph W. Zarzynski)

80s degrees Fahrenheit. This was certainly not the type of weather conducive to distance running, especially an ultramarathon over a hilly course. As our sedan cruised along General Wade's Military Road, I sure was hoping the weather would cool down for my ultra.

After our visit to historic Culloden and before we got to Fort Augustus, we spent some time at a pull off alongside Loch Ness on the east side of the waterway. At that uncrowded site, we conducted shore watching with our binoculars and cameras, scanning the waters of Loch Ness hoping a Nessie would pop to the surface. Occasionally, we would notice other monster enthusiasts with their cameras.

After we completed the shore watching, we motored to Fort Augustus and there we ate an early dinner. After our meal we drove to Adrian Shine's camp near the canal in Fort Augustus. Adrian was glad to see us, and we had a congenial conversation about Nessie. He showed us a mammoth trap in a field that he had been commissioned to fabricate. The construction job supported the scientific work of the Loch Ness Project, the new name for his research group. In the year 1981, the "Project" acquired sonar equipment, an echo sounder, for their boat work which they employed for several years on the loch (Harmsworth 1984). Shine's fabrication of the Nessie net helped fund their scientific efforts at the waterway. This so-called "monster trap" was financed by a liquor company that wanted to use their hunt for Nessie to promote their beverage spirits.

The entrapment scheme, approved by local authorities, was funded by Vladivar Vodka. The cylindrical fiberglass net, that I estimated at approximately 60 x 20 x 20

9.6 This cylindrical-Nessie trap lies in a field near Loch Ness. The voluminous net, funded by a British-vodka company in 1984, was suspended in the waterway in a bold attempt to capture a Nessie. (Credit: Pat Meaney)

feet, was eventually lowered into Loch Ness. The objective was to capture a Nessie animal. In the attempt to do so, the "Project" was sure to acquire loads of publicity for the alcohol firm. Just days after Pat and I departed Loch Ness in 1984, a helicopter transported the device to the waterway. The jumbo trap was lowered into the water near the shore at a landmark known as the "Horseshoe Scree." This site, on the east side of the loch, was a geological irregularity, a scar of rocks, horseshoe-shaped. According to Highland lore, "Horseshoe Scree" was believed to be the spot where a gargantuan Nessie once pulled itself up onto shore. The unusual netting reportedly remained in Loch Ness for several weeks, but no Nessie was cap-

9.7 "Horseshoe Scree" is a huge horseshoe-shaped scar of rocks found on the east side of Loch Ness. According to local lore, "Horseshoe Scree" is the location where a gargantuan Nessie once crawled up onto shore. (Credit: Martin Klein)

tured (Zarzynski 1986:65-66). The goal of most cryptozoologists of that era was to prove the existence of these unknown animals through the acquisition of photographs, movie film, video, or recovering a carcass from a Nessie animal that died a natural death. It was not to capture or kill a Nessie simply to prove its existence.

In 1987, Adrian Shine would be one of the principals in Operation Deepscan, a highly-publicized Anglo-American sonar survey of the loch. Utilizing 24 vessels, each equipped with a Lowrance X-16 sonar, the crafts were strung out in a row across the width of the loch. Lowrance was a Tulsa, Oklahoma-based consumer-sonar company that principally manufactured fish-finder sonar units. The flotilla "mowed" a lengthy section of the waterway hoping the survey operation would pick up a sonar record of a Nessie creature. Frequent strong winds hindered much of the project, yet three-sonar contacts were obtained. However, the results were not definitive (*The Telegraph* 2016).

August 20th had certainly been a busy day for us. First, Pat and I visited Jim Hogan near Inverness. We then toured Fort George and Culloden Battlefield, did shore watching hoping to photograph a Nessie, visited monster hunter Adrian Shine, and inspected and photographed Adrian's handiwork of the monster trap. If Pat and I kept up this hectic pace, I would be too fatigued to "run with Nessie" in only 36 hours.

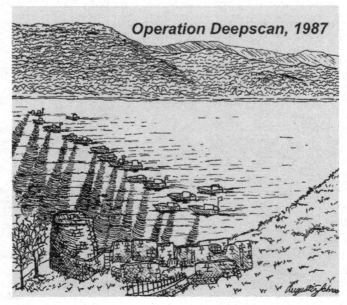

9.8 An artist interpretation of Operation Deepscan, a 1987 project that employed a flotilla of vessels, each equipped with Lowrance hull-mounted sonar. The group hoped to solve the Nessie mystery, but the project ran into poor weather. (Credit: August Johnson)

Tuesday—August 21, 1984:

Following breakfast at Benleva Hotel, Pat and I again visited Tony Harmsworth at his Drumnadrochit office. While there, Erik Spence from Moray Firth Radio in Inverness, telephoned Tony to inquire of any updates on what was new in the monster-hunting field.

Moray Firth Radio was relatively new, only a couple of years in operation. Spence had likewise contacted Harmsworth to ask about another topic, my upcoming ultrarun of Loch Ness. The air-waves personality found out about my solo run from a news release sent to him from Harmsworth. Tony was superb at keeping the media apprised of interesting and unique events related to Loch Ness. After a brief conversation with Spence, the radio personality invited me to come to the station to do a radio interview about my ultramarathon. I eagerly agreed. Pat and I promptly drove to the radio station where we were greeted by the enthusiastic and cordial Erik Spence. I did a 12-minute interview that was scheduled to be played the next day, beginning in the early morning.

Spence was a talented interviewer and he seemed genuinely intrigued with the run. The radio host believed it had a human-interest component and because it was a local happening, it was newsworthy for Moray Firth Radio's listeners. Spence said the radio station would also do periodic announcements to inform drivers to watch out for a tall runner bounding along route A82. I realized, maybe for the first time, that I was probably perceived as not just a monster hunter, but also as a peculiar American obsessed with pounding the pavement along busy-route A82.

The concept of someone running the length of Loch Ness was at that time quite "foreign" to many Highlanders. Why would anyone want to run on that busy stretch of A82? Likewise, over my past few visits to Scotland and during my training jogs in Drumnachochit and Inverness, I seldom saw another runner.

My ultramarathon along the two-lane vehicular artery could be a hazard for drivers, and of course, for myself, the eccentric "Yank." The radio briefs announcing the upcoming event would promote safe driving and also would provide some moral support from passing motorists. The thoroughfare had minimal shoulders adjacent to the paved road. I would be running on the right side of the road, heading into oncoming traffic. I knew I would have to frequently dodge cars, vans, and lorries by leaping off the road, often over a metal guardrail, or up onto a short stone-wall that ran along sections of the vehicular lane. To add, for parts of the stretch of Loch Ness, route A82 was elevated more than a 100 feet above the water. That would be a long and possibly fatal fall if I tumbled off the side of the roadway.

Following the radio interview with Erik Spence, Pat and I strolled around the streets of resplendent Inverness, especially busy High Street filled with its shops and eateries. The Scottish Highlands capital had many attractive stores, especially those that sold woolen garments produced in Scottish mills. We also visited the city's library and did research reviewing back issues of the *Inverness Courier* to see if there were recent sightings of Nessie reported in that local newspaper. After our archival exploration, we had a late lunch and returned to the Benleva Hotel to rest, since my ultrarun was the following morning.

After a quiet afternoon in the hotel, we motored down to Fort Augustus to again converse with Adrian Shine. I was quite impressed that a person with such a demanding schedule as his would allot that much time to a question-and-answer session. Following a congenial meeting with Adrian, Pat and I ate dinner at the Loch Ness Lodge Hotel in the center of Drumnadrochit. Then we returned to our hotel room for what turned out to be a tranquil sleep for both of us, the ideal prep for my distance run the following day. Before sleeping I prayed that a mid-week scamper over the asphalt of highway A82 would see a low volume of traffic and with optimal weather, too. I likewise hoped Pat would be well-rested, since she would play an integral role as my support team.

Chapter 10

My One-Person Loch Ness
Ultramarathon—Just Before the Big Run

Wednesday—August 22, 1984

It was finally "ultrarun day." My fiancée and I got up early, at 5:45 a.m. I would later write in my journal: "It was a wee bit cooler than the last couple of mornings." This was indeed fortunate because I had earlier in my running career had two bad experiences after long-distance races, both requiring medical attention due to heat, humidity, and dehydration. Years later in 2019 and 2020, I would be examined by multiple doctors and also would take a battery of medical tests trying to ascertain why I often suffered from humidity-related issues during and immediately after running events. So, I was elated that August 22, 1984 was not a hot morning.

The first time I was afflicted with dehydration in a distance race was at the Marine Corps Marathon in the Washington, DC on November 1, 1981. I was 31 years old, in peak physical condition, and it was my second marathon. I was still, nonetheless, a novice at distance running and this marathon was Pat's only 26.2-mile race. She had trained with me for that 1981 marathon.

Years later, Pat would remind me of one jog in particular. It was in June 1981 and we were staying for a couple of weeks at Lake Champlain in a lakeside cabin we had rented near Vergennes, Vermont. Following our cryptozoological fieldwork for the day, we drove the 30 miles to the city of Burlington, Vermont to have dinner at a restaurant. We also brought our running clothes and shoes. It was evening when we were driving back to our cabin. I suddenly realized we did not get in a run that day.

10.1 Pat Meaney and the author in the early 1980s based at a lakeside cabin searching for the Lake Champlain monsters. (Credit: Joseph W. Zarzynski Collection)

I pulled my cargo van off the road and suggested to Pat that we should get dressed into our running garb for a short jog. Pat thought I had forgotten about running that day. She became a little upset that so late in the evening, after 9:00 p.m., that I desired to go jogging. It was not so much that we needed the exercise since both of us were in excellent physical shape. It was, however, purely honing the mental edge. To this day, Pat still occasionally reminds me of that training run, calling it—silly, ridiculous, not worth it, and other unflattering words.

A few months later on November 1, 1981, it was the Marine Corps Marathon, a race around Washington, DC and parts of northern Virginia. I finished minutes ahead of Pat. As I was waiting for her to complete the marathon, I began to feel lightheaded and overly fatigued. I was later told that I had collapsed at the crest of grassy hill near the finish line that was not far from the US Marine Corps War Memorial (aka Iwo Jima memorial) in Arlington, Virginia. I don't remember falling down. What I do recall was waking up in a military medical tent with Marine Corps doctors, nurses, and medics attending to several runners, including myself. Most of us had medical issues stemming from the day's heat and humidity. I must admit, I was embarrassed that I required health care. I thought I drank enough water in the couple of days prior to my marathon. My treatment was for "extreme dehydration," that resulted in me passing out. After all, I had just run a marathon in humid weather. Upon being revived, I realized I had an IV in not one, but both arms. An IV is a needle stuck in a patient's vein with fluids being administered via a tube and bag. The therapeutic pouch generally contains salt, sugar, electrolytes, and even vitamins. Having two IVs, rather than just one, was a telltale indication that I was actually in rough shape.

After Pat finished the race, she finally tracked me down. Pat had been unaware of my calamity since she finished the race minutes behind me. Before I was released from medical treatment, I remember hearing over the event's loudspeaker—"Will Joseph Zarzynski report to the finish line." Before I could speak, another fallen athlete near me, who likewise suffered from dehydration, shouted, "I'm Joseph Zarzynski." I immediately countered in an ear-splitting voice, "No, I'm Joseph Zarzynski." After nearly two hours in the Marine Corps emergency tent, the medical staff finally released me. Pat was satisfied that I was okay. I felt much better though my feet and legs were sore from the 26.2 miles of pounding the pavement.

It was a long and painful day, but we both finished the marathon. Pat was so excited that she had become a "marathoner." I was extremely proud of her accomplishment, too. I completed the event in 3:49:00 (gun time). Pat's time was quite respectable, too, for a first-time marathoner. She finished the race in 4:42:05 (gun time) and Pat had no problems during or afterwards, unlike her companion.

My second experience with heat and dehydration was in 1983, during the Sybil Ludington 50-kilometer run. It was held in Putnam County in the Taconic Mountains of downstate New York. This was my first ultramarathon and it was a demanding run, too, 31.1 miles over a demanding course. Because of the distance and the hilly terrain, this was one of the most difficult races of my life.

I selected that 1983 ultrarun because I love history and the race overflowed with colonial-American heritage. As a social studies teacher I was well-versed in Revolutionary War (1775–1783) history, even the little-known story of heroine

Sybil Ludington. The run was named after Ludington, often dubbed "the female Paul Revere." On April 26, 1777, Sybil Ludington, only 16 years old, rode a horse from village-to-village in Putnam County, New York and the surrounding district, a circuit of about 40 miles. Her gallop notified local patriot militias that a British army was approaching. Though the British redcoats succeeded in their raid upon Danbury, Connecticut, the American patriots soon met the enemy and pushed the British back toward Long Island Sound (American Battlefield Trust 2020). Like many heroes of America's past, we sometimes don't know much about them or their courageous exploits until well after these people died. Such was the case with Sybil Ludington. Decades later a statue of the Revolutionary War heroine was erected in Carmel, New York. A plaque at the monument reads:

> "Sybil Ludington — Revolutionary War Heroine April 26, 1777. Called out the volunteer militia, by riding through the night, alone, on horse-back, at the age of 16, alerting the countryside to the burning of Danbury, Conn. by the British." (*History Is Now Magazine* 2019)

The Sybil Ludington Historical Run in 1983 was on April 30. I completed the 31.1 miles in a pedestrian pace, 4 hours, 51 minutes. The abundance of steep hills contributed mightily to my slower pace. As well, it was the first time I had run over 26.2 miles, so I didn't push my pace.

After the ultra, I sipped some water, hugged my girlfriend, and we slowly walked to the post-race celebration at a nearby restaurant. I do remember strolling over to the bar from our dining table at the meal to get a much-anticipated lager. I secured the chilled beer in my right hand, and then I promptly passed out with my head deposited on the wooden bar. It was a poor optic for the race contestants and their families. After a harrowing ambulance ride, Pat and I arrived at a hospital in nearby Mahopac, New York. I was ushered into the emergency room where I was diagnosed with a case of acute dehydration. Once again, I was given an IV and fruit juice to drink. After a few minutes I began to recover from that frightening experi-ence, one that terrified Pat. It was after all, my second medical treatment following a distance run. I was now a veteran of "runner down." Eventually, I was released by the doctor and nurses. Pat was forced to drive my van back to our apartment. Of course, I slept most of the three hours home. It was a slumber interrupted several times by Pat, inquiring if I was still alive.

Nearly a year-and-a-half after my Sybil Ludington 50-kilometer run, I was about to undertake another ultramarathon. I was very cognizant during the days leading up to the Loch Ness run to drink plenty of water and other fluids to properly hydrate.

In the early morning hours of August 22, 1984, Pat and I motored down route A82 to a self-imposed location in the heart of Fort Augustus. The starting line was well behind, that is, south of the head of Loch Ness. I selected that spot to ensure there would be no arguments that the starting line was not south of Loch Ness. It was already a warm morning for the Scottish Highlands, but the weather forecast at least did not call for blistering heat like earlier in the week.

On ultrarun day, I wore a light-colored T-shirt with a logo of a runner from my 1982 Skylon International Marathon, a course from Buffalo, New York to Niagara Falls, Ontario, Canada. I was in light-blue shorts, white athletic socks, and a stretch headband to keep the sweat from my eyes. I likewise had a Casio wristwatch to time my ultrarun. I did not wear my "new" spectacles, the cat-eye sunglasses I purchased several days earlier in a flea market in Glasgow. They literally could have caused a vehicle-runner accident. If the opportunity arose while I was running, I hoped to spend a little time Nessie watching, perusing the surface waters hoping to catch a glimpse of the "hidden ones." Though I needed eyeglasses to see the text of some distant road signs, I was nonetheless hoping if a 20-to 30-foot, dark-colored creature surfaced on Loch Ness, it would be visible if I squinted hard enough.

Just before the start of the run, I hugged my lovely and supportive fiancée. We then wished one another good luck in our dual endeavors—her driving beside the loch and me running my one-person ultramarathon.

Chapter 11

My One-Person Loch Ness Ultramarathon—
Fort Augustus to Invermoriston

. . . .

Wednesday—August 22, 1984 (continued)

Just before the run started, Pat read a biblical quote, a tradition she did for my distance races since our Marine Corps Marathon in 1981. The passage was from Isaiah 40: "…they shall mount up with wings as eagles, they shall run, and not be weary." This was also a line in the Oscar-winning movie—*Chariots of Fire* (1981)—a British film that was beloved by runners all across the world. My one-person run with Nessie, the Loch Ness water horses, began at 7:03 a.m.

I was not worried about myself. Later that day, after the run, I recorded in my journal my feelings just before the ultramarathon. I scrawled: "I was very concerned about her [Pat's] safety in the car driving, but I knew she could do it…."

I was running on the right-hand side of route A82, facing oncoming traffic. Pat had driven ahead to find the first place to turn off to wait for me so she could give me water. Years later, Pat informed me she still sometimes recalls pulling off the road that initial time, but not at a lay by. Some A82 motorists were upset that her sedan was not parked in the safety of a vehicle lay by, so they began honking their car horns. Pat reflected, "I quickly learned only to pull into a lay by that was safer for all" (Meaney 2019-2020).

I recollect that shortly into my ultrarun, route A82 briefly swerved away from the waterway. Soon I was rushing past the stately Inchnacardoch Hotel and thinking, what a majestic building. The impressive stone structure had been the hunting

74

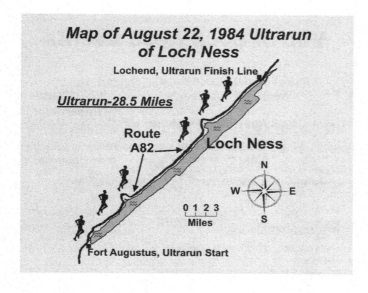

11.1 Map showing the route of Joseph W. Zarzynski's August 22, 1984 solo ultrarun of Loch Ness. (Credit: Joseph W. Zarzynski)

lodge of the Lord Lovat family. The first Lord Lovat was a fifteenth-century leader of the Fraser Clan of Lovat. Older readers of this book might recall the 1962-block-buster movie, *The Longest Day*, about "Operation Overlord," the June 6, 1944 Allied invasion of Normandy, France. The Oscar-nominated, black-and-white classic film, produced by Hollywood-kingpin Darryl F. Zanuck, cost a hefty $10 million to produce (Denby 2014). That was a lot of money at that time. The flick had an international ensemble of actors including Peter Lawford, a British actor who was also the brother-in-law of President John F. Kennedy. Lawford portrayed the 15th Lord Lovat (aka Simon Fraser), a brigadier general and distinguished commando in the British Army during World War II.

Earlier in the week, Pat and I had eaten dinner at the historic Inchnacardoch Hotel. As I trotted past the inn, I thought how we were so impressed with not only the regal-looking dwelling, but also our delightful meal. The building is set back off route A82, elevated atop a slight knoll overlooking the southwest corner of Loch Ness.

The imposing lodge is situated across from Cherry Island on Loch Ness, a manmade island that was once a crannog. This crannog was an ancient timber-built and fortified artificial island, somewhat circular in shape. It lies in Inchnacardoch Bay, about 150 yards from shore. Cherry Island is today the only island on Loch Ness and thus it is frequently photographed by boaters and others. When the

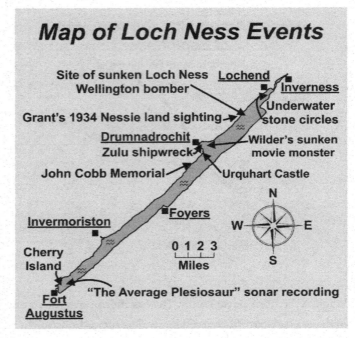

Map of Loch Ness Events

Site of sunken Loch Ness **Lochend**
Wellington bomber
Inverness
Underwater
Grant's 1934 Nessie land sighting → stone circles

Drumnadrochit
Zulu shipwreck
Wilder's sunken movie monster

John Cobb Memorial
Urquhart Castle

N

Foyers
Invermoriston
W — E

0 1 2 3
**Cherry
Island**
Miles
S

"The Average Plesiosaur" sonar recording
**Fort
Augustus**

*11.2 Map show-
ing the locations
of some of the key
events associated
with Loch Ness.
(Credit: Joseph W.
Zarzynski)*

Caledonian Canal was built in the early 1800s, the newly constructed waterway raised the loch's water level by about six feet. Cherry Island in 1984, had several trees growing on it and I'd estimate the islet to be less than 100 feet in diameter.

In 1908, when the water level of the loch was down, a diving survey was conducted around the crannog by none other than a Catholic cleric named Reverend Odo Blundell. He wanted to determine if the oblong-shaped isle was indeed an artificial island. Blundell was a novice diver whose subsurface investigation was undertaken using hardhat-diving equipment loaned to him with surface-supplied air pumped down via a hose. Blundell explored the loch bottom all around the small island making observations and probing to uncover the origins of Cherry Island. The intrepid underwater surveyor concluded that Cherry Island was indeed artificial and that it was an ancient crannog. The island apparently once included a causeway that headed from the crannog toward three boulders on the mainland. Furthermore, before the canal raised the water level in the loch by about six feet, Cherry Island was much broader. Decades ago, the island was sometimes known as *Eilean*

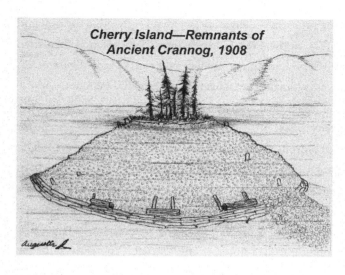

Cherry Island—Remnants of
Ancient Crannog, 1908

11.3 An artist's inter-pretation of a 1908 drawing by Reverend Odo Blundell of the remains of the Cherry Island cran-nog. Located at the south end of Loch Ness, this was once an ancient-artificial island. (Credit: August Johnson)

Muireach (Murdoch's Island). Reportedly, earlier in history there was another near-by isle, smaller in size, known as *Eileen Nan Con* or Dog Island. However, with the raising of the water due to the Caledonian Canal's construction, Dog Island was engulfed (Blundell 1909:159-163).

In 1984, Cherry Island was a one-of-a-kind exquisite sight. The tiny island only had a few trees on it, but it stood like a defiant sentry against the vastness of Scotland's most impressive waterway. As I jogged by, I thought how I wanted to one day take Pat in one of the Caley Cruisers' boats and anchor off the crannog to inspect this pretty little piece of real estate.

Running along the western shore of the loch, I was struck by the forestlands to my left along route A82. Some of hillside vegetation appeared to have been recently cut down, too. Today, I understand the vegetation on both sides of route A82 along the waterway has grown up substantially and in some cases obstructs the view of the imposing loch.

At about 4 ½ miles into my run, I darted behind some of that formidable tim-berland to relieve myself, my first of three bathroom stops during the 28.5-mile ultra. Many veteran marathoners will tell you that urinating during a race is a not a bad thing. During each stop, you might add half a minute or more onto your run-ning time, but it means your kidneys are properly functioning. Beginning a cou-ple of days before race day, I started consuming more water than unusual. I was not going to have the same problem—dehydration—that I suffered in the heat and humidity of Washington, DC at the 1981 Marine Corps Marathon and then later, after completing the 1983 Sybil Ludington 50-kilometer run. I drank plenty of fluids, so during my Loch Ness ultrarun, the obligatory peeing I did behind trees.

What was spectacular about the landscape around Loch Ness was that for the most part, route A82 pretty much hugged the shoreline of Loch Ness except for two meandering forays away from the iconic waters. These diversions were first at Invermoriston and later at Drumnadrochit. This meant, though my utmost concern was keeping an eye out for potentially dangerous vehicular traffic, at times I had a decent view of the loch's surface and from an optimal vantage-point, too, an elevated roadway. I really was "monster hunting on the run." It was an inspiring feeling.

That evening, after my ultrarun, I noted in my journal, "The early ⅓ of the route was climbing uphill. I really didn't mind that." Well, that was because I was in terrific running shape. I had trained on a lot of rolling terrain over the avenue of North Broadway and the Skidmore College loop, both in Saratoga Springs, as well as frequent jogs along the undulating lanes of the Saratoga Spa State Park, my primary training grounds. That hill climb from Fort Augustus toward Drumnadrochit often provided me with awesome views of the water. Whenever possible, I took any opportunity to scrutinize that magnificent scenery.

As I charged ahead, I was extremely grateful for the radio interview I conducted the previous day at Moray Firth Radio. Erik Spence and his colleagues at the station announced occasional communiqués over their air waves during the morning of August 22, 1984, reminding motorists on A82 to beware of the solo runner. The Moray Firth Radio alerts worked, too, as most vehicular travelers smiled and many even waved. There were very few who were visibly perturbed that I was sharing "their" road. For the most part, the local populace was wonderfully supportive. A couple of times I said out loud to myself, "Wow, these Scots are terrific drivers."

Borrowing a quote from British mountain climber George Herbert Leigh Mallory when asked in 1923 why Andrew Irvine, his climbing colleague, and he were attempting to scale Mount Everest (29,032 feet in elevation), Mallory exclaimed, "Because it's there." Well, I felt somewhat the same way about my Loch Ness run. A couple of times I reflected upon Mallory and Irvine's assault on "Everest." Those undaunted climbers died during their 1924 attempt to be the

11.4 The author running in the morning haze during his August 22, 1984 ultra marathon at Loch Ness. (Credit: Pat Meaney)

first to scale the tallest peak in the world. I certainly was not in the same rarefied air as those adventurers, nor did I want a similar fate. Still, a few vehicles buzzed pretty close to me and I murmured, "Zarzynski, what the hell have you gotten yourself into?"

As I "pounded the pavement," Pat drove the automobile, finding safe spots to provide water, give verbal encouragement, and even snap a few photographs to document my run. Pat was a first-rate cheerleader and I am to this day surprised she didn't smuggle a cow bell into our luggage to complement cheering sessions.

From the very start of the run, I counted vehicles that passed me going in the same lane. An archaeologist friend of mine, Dr. David Starbuck, who had taken groups of college students and others on cultural tours around Scotland, once informed me that the traffic nowadays along route A82 would be too busy for a run like what I did in 1984. I guess my one-person ultra in 1984 was indeed timely. A couple of times I spotted a police car on the route, but they did not seem too perturbed. I was lucky there.

Sometimes, especially when wide lorries and buses were heading my way, I was forced to jump over guardrails (aka crash barriers) that were positioned along route A82, a vertical leap of about two feet. Those frequent pauses during my ultra-marathon certainly added many minutes to my ultrarun's time. I guess my years of playing basketball and jumping for rebounds paid off, though I was never known as a hoops player with great "hops."

As I approached about six miles into the ultrarun, route A82 swung to my left toward Invermoriston. The village was less than a mile from the loch and Invermoriston was renowned for its famous bridge. This was the site of the historic stone-span over River Moriston, erected in 1813 under the direction of Scottish engineer Thomas Telford. He likewise managed the construction of the Caledonian Canal. In the 1930s, another bridge for the newly constructed route A82 was built in Invermoriston. I believe that was what I crossed on my way toward the next stop in my solo trek, the spacious parking lot at Urquhart Castle. That loch-side fortress was a few miles from Invermoriston and not far from the village of Drumnadrochit.

Chapter 12

My One-Person Loch Ness Ultramarathon—
Invermoriston to the John Cobb Memorial

Wednesday—August 22, 1984 (continued)

At Invermoriston, route A82 turned back toward the shoreline of Loch Ness. Further, at Invermoriston, route A887 from the west, entered into route A82. When I saw the sign for A887, I thought back to my 1979 trip to Loch Ness.

I had come to Loch Ness in the summer of 1979 for two reasons. First, because the Academy of Applied Science planned to use a pair of dolphins in the loch that were each to have been equipped with a single underwater-camera to attempt to photograph a Nessie. Each camera weighed about four pounds. The dolphins had undergone extensive training in Florida, teaching them to swim into their camera harness, a design that was supposed to be minimally intrusive for the mammals. Moreover, the dolphins had experienced gradual-acclamation forays from saltwater into freshwater, like the freshwater of Loch Ness. Thus, for short time-periods, the AAS hierarchy believed the dolphins could undertake swims in the loch. The mammals' innate sonar might have been able to locate a Nessie creature and photograph it (Wilford 1979). Sadly, in June 1979, the Associated Press reported that one of the dolphins had died in a holding pen awaiting its plane flight to Scotland. Thus, this project was cancelled (Associated Press 1979 and Rines 1979).

The U.S. Navy Marine Mammal Program has been utilizing trained dolphins since at least 1959, mainly because of their acute echolocation abilities. Not only

12.1 One of the Academy of Applied Science's camera-equipped dolphins. In 1979, the mammals were trained to use their echolocation to swim up to a Nessie animal and photograph it. The planned operation at Loch Ness was never undertaken. (Credit: Martin Klein)

were these aquatic mammals able to engage their incredible sonar to detect mines lying on a seafloor, they also could discourage enemy divers and discover other subsurface military threats. More recently, it has been determined that dolphins can locate three-inch diameter balls from a distance of 584 feet (Gilliland 2019). It was no wonder that the AAS team hoped that trained dolphins, searching for a limited number of minutes in the depths of Loch Ness, could detect and photograph a swimming Nessie.

The second reason I traveled to Loch Ness in the summer of 1979 was to do fieldwork with a friend, cryptozoologist Tony Healy. The Australian was on a two-year-long global tour visiting sites where there might be hidden animals. Tony had already explored parts of Canada and the USA searching for Bigfoot creatures. He likewise visited Lake Champlain to look for Champ. Before his monster-tracking excursion ended, Tony would visit Ireland to examine tales of some of its mysteries and he spent a few months at Loch Ness, too.

Tony Healy was what I would describe as a minimalist, since at Loch Ness he lived out of a light-blue Ford Escort van. Tony camped at lay by spots around the

12.2 Tony Healy, an Australian cryptozoologist, at Fort Augustus, Loch Ness in 1979. (Credit: Joseph W. Zarzynski)

loch, conducting shore watching with camera and binoculars. Months earlier, Tony spent several days sleeping on an uncomfortable couch in my bungalow residence in Saratoga County. We quickly became buddies. I appreciated his independence and his enthusiasm for investigating unknown animals and other phenomena. Years later, Tony would co-author (with Paul Cropper) three books on his numerous investigations: *Out of the Shadows: Mystery Animals of Australia* (1994), *The Yowie: A Search for Australia's Bigfoot* (2006), and *Australian Poltergeist: The Stone-throwing Spook of Humpty Doo and Many Other Cases* (2014).

While I was at Loch Ness in July 1979, Tony and I spent 12 days undertaking camera watches at various locales around the loch. It was wonderful fun and "for the cause." We likewise visited with several cryptozoologists that were at the loch that summer—Tim Dinsdale, Dr. Robert Rines, Dick Raynor, Ivor Newby, Charlie Wyckoff, Holly Arnold, and others. These were among the "Who's Who" of Nessie detectives. Further, while in Scotland during that summer, Tim Dinsdale, known to many as "Mr. Loch Ness," inducted me into his "Loch Ness Association of Explorers." It was a society of a handful of Nessie hunters and I was quite honored to become a member.

During that 1979 expedition to Loch Ness, Tony and I took a break one day from our fieldwork and we visited a hamlet on route A887 that was called Dundreggan. It was just down the road from Invermoriston. Dundreggan lies in Glen Moriston (aka Glenmoriston) on the north side of River Moriston. I first read about Dundreggan in Nicholas Witchell's superbly researched and written book—*The Loch Ness Story* (1974). Witchell had been a Nessie hunter as a young man and later he became a celebrated television-reporter and newscaster with the BBC. He and I exchanged several letters over the years about our mutual interest in Nessie. Witchell's book on Nessie had a one-paragraph description of Dundreggan (Witchell 1974:37). Since the word—Dundreggan—means "Dragon Haugh" or "hill of the dragon," it was naturally a locale that Tony Healy and I had to investigate.

As I was trotting along toward Invermoriston during my ultrarun, my mind was filled with pleasant memories of that 1979 sojourn that Healy and I took to Dundreggan. It was July 14, and we drove from Loch Ness over to route A887 and to Dundreggan, located west of Invermoriston. The site was in an area known for its birch-juniper woodlands. We were in search of the "hill of the dragon" which according to local lore was called the Dundreggan Fairy Knoll. One local legend was that centuries ago the warrior Fingal slew a fierce dragon. The hero buried the monster right where it fell. Thus, over time, the grave became a grassy knoll, partially covered with rocks. The tale goes that little fairies took up residence in this mound. An ancient warrior, a dragon, and pixies, what more could you ask for inquisitive Nessie hunters?

When Tony and I arrived in Dundreggan, we asked around and were given directions to the "hill of the dragon." We drove to where the site was said to be located and spied a nearby house. Adhering to the phrase—*carpe diem*—we decided to knock on the cottage door, explain to the occupant the reason for our visit, and seek information about the legend of Fingal and the dragon. An elderly lady answered the door. She was extremely pleasant and invited us in for tea. Not only were we treated to a couple of cups of tea and several biscuits, she entertained us with incredible stories about Dundreggan, as we also watched some sports on her "telly." After saying good-bye to this gracious person, Tony and I sauntered over and viewed the fabled "home of the dragon." We snapped several photographs of the storied hillock. I would visit Dundreggan again, with Pat on August 11, 1982.

After passing through Invermoriston, route A82 headed back toward Loch Ness. When the roadway was again running alongside the waterway, there was then

12. 3 Dundreggan's "hill of the dragon." Folklore suggests this is the grave site of a dragon killed centuries ago by an ancient warrior. The hillock is six-miles west of Loch Ness. (Credit: Joseph W. Zarzynski)

a long-straight stretch for a few miles. I was just across the loch from the village of Foyers, located on the eastern shore of the waterway. Foyers, on B852 roadway (General Wade's Military Road), was the site of a hydroelectric plant and also the Falls of Foyers. Two years earlier in August 1982, Pat and I stayed a night at the Foyers Hotel.

A few miles further into the ultramarathon was Achnahannet, a tiny hamlet on my way toward historic Urquhart Castle. Nessie aficionados will tell you that Achnahannet was the site of the headquarters of the Loch Ness Phenomena Investigation Bureau (LNPIB). This organization was formed in the early 1960s, principally by David James, a member of Parliament (Bauer 1986:86).

As I approached Achnahannet, I thought back to a visit Pat and I had with David James on August 14, 1982. We met him at Torosay Castle, his residence on the Isle of Mull in the Scottish Inner Hebrides. James had a remarkable life and was a fervid supporter of the search to find Nessie. We had lunch with him at his home while he entertained us with accounts of his service in the Second World War and later when he led the way founding the LNPIB.

David James was a bonafide naval hero. During the Second World War in 1943, while in the British Royal Navy, his patrol vessel was sunk by Germans. He was captured and sent to a Nazi prisoner-of-war (POW) camp for Allied naval personnel. Later that year, posing as a Bulgarian naval officer, James escaped from the POW

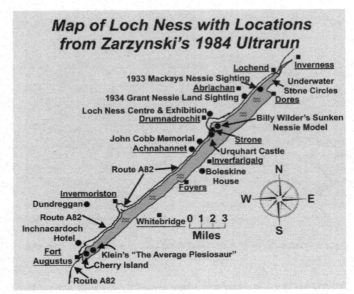

Map of Loch Ness with Locations from Zarzynski's 1984 Ultrarun

12.4 Map of Loch Ness with Locations from Zarzynski's 1984 Ultrarun. (Credit: Joseph W. Zarzynski)

facility (Witchell 1974:153). However, he was recaptured and sent back to a POW camp. In 1944, James again broke out of his confinement. His second attempt at freedom succeeded. He had disguised himself as a Swedish sailor and after days on the run, James finally arrived in Stockholm, Sweden and safety in that neutral nation. For his wartime exploits, he was awarded a Distinguished Service Cross and later received the Order of the British Empire (OBE). After the war, David James wrote a well-received book called *Escaper's Progress*. The publication was about his noteworthy military service that included his two daring escapes from German POW camps (James 1978). James later served as a member of Parliament in the House of Commons. In the early 1960s, with encouragement and support from Sir Peter Scott, a British naturalist, James started the LNPIB (Witchell 1974:155). Sadly, in 1986, just a few years after our visit with David James at Torosay Castle, this decorated military veteran and former Parliament member passed away (The Guardian 1986).

Besides conducting shore watching and some fieldwork from vessels on the loch, LNPIB members also erected a modest and temporary visitor-center at their headquarters at Achnahannet. The LNPIB attracted a wide variety of international volunteers, one being Dr. Roy P. Mackal. The Wisconsin-born Mackal was a former-U.S. Marine who later became a biochemist/zoology professor at the University of Chicago (Coleman 2013). Other LNPIB members included diehard monster hunters like Tim Dinsdale, fascinated college students, and explorer-type people, all intrigued by science, adventure, and mysteries. The LNPIB existed for over a decade from the early 1960s to about 1972. Most of their fieldwork was shore-based camera surveillance. They employed cumbersome, but powerful cameras that were mounted upon tower platforms and also on scaffolding erected atop vans. These vehicles were driven to strategic locations around the loch so that the long-range cameras could provide as much coverage (eyes and cameras peering out over the waters) as was possible (Mackal 1976:17-19). According to Dr. Henry H. Bauer, a professor of chemistry and other sciences at Virginia Polytechnic Institute and State University (aka Virginia Tech) in Blacksburg, Virginia and the author of the book *The Enigma of Loch Ness: Making Sense of a Mystery* (1986), at times this

12.5 *In 1968, this camera platform erected upon a van was one of several stationed around Loch Ness by the Loch Ness Investigation (aka Loch Ness Phenomena Investigation Bureau) in their multi-year search for Nessie. (Credit: Max Scheler/SZ Photo and Alamy)*

camera surveillance covered more than 70 percent of Loch Ness (Bauer 1986:86).

Visitation to the LNPIB's Nessie exhibit at Achnahannet went from a mere 1,000 visitors in 1965, to 25,000 people in 1966, and then more than 54,000 during 1971 (Bauer 1986:88). Clearly, by the mid-1960s and into the 1970s, there was a robust interest in Nessie-generated cultural and scientific tourism at the loch. Nick Witchell later wrote that the Loch Ness Phenomena Investigation Bureau "closed down at the end of 1972 through lack of funds and the necessary planning approval from Inverness-shire County Council to continue occupation of the H.Q. [head-quarters] site" (Witchell 1974:195).

After I passed Achnahannet that morning, I was roughly at the midway point into my ultrarun. Regrettably, I saw no signs marking the site of the once-busy encampment and information center. For me, it was onward to the John Cobb Memorial and then Urquhart Castle, the latter a stone stronghold located just south of Drumnadrochit.

Chapter 13

My One-Person Loch Ness Ultramarathon—
John Cobb Memorial to Beyond Urquhart Bay

Wednesday—August 22, 1984 (continued)

On my way to Urquhart Castle, I passed the John Cobb cairn, a memorial mound of stones with a bronze plaque. I was at the 15 ½-mile point into my run. The monument is located about 1 ½ miles from the castle ruins. John Rhodes Cobb was a former pilot in the Royal Air Force and later a successful fur broker for garments. He later became a world-class automobile racer and powerboat pilot. In 1947, Cobb set a speed record on land of 394.2 mph. The English speedster then turned his attention to setting records upon water. His vessel, the *Crusader*, was built over the years

1949 and 1952, and the craft was powered by a jet engine. It was probably the first speedboat with that type of powerful propulsion.

13.1 The John Cobb Memorial with a bronze plaque mounted on a stone cairn. The monument honors speedboat-pilot John Cobb, who died at Loch Ness during his attempt in 1952 to set the world-speed record on water. (Credit: Joseph W. Zarzynski)

In the summer of 1952, Cobb ventured to Loch Ness eager to smash the world-speed record on water that was 178.497 mph. Following days of practice runs, on September 29, 1952, John Cobb was ready to try to establish a new speed record on water, using legendary Loch Ness for that attempt. Cobb was required to make two runs in his vessel, each over a measured mile. The average of the two sprints would be his recorded speed.

Cobb's first trip in his swift watercraft was recorded at 206.890 mph. His second attempt, reported to have attained speeds up to 240 mph, skipped along until it then appeared his watercraft hit three ripples. That surface disturbance caused the powerboat's nose to suddenly dip. The *Crusader* then veered to the port and broke apart.

The 52-year-old John Rhodes Cobb died in that terrible crash. His body was recovered, but his aluminum-and-wooden speedboat sank into the recesses of the loch. He most likely had gone faster on water than any other person to date. However, because he did not complete the second trip, he was not posthumously awarded a world record (Zarzynski 1986:6-9). After his tragic death, a cairn was

13.2 John Cobb's-Crusader powerboat on Loch Ness in 1952. Cobb was killed during an attempt to set the world-speed record on water when his craft hit waves and broke apart. (Credit: J.A. Menzies)

erected as a cenotaph along route A82 near where the accident occurred. A bronze plaque on the cairn reads:

On the waters of Loch Ness John Cobb having travelled at 206 miles per hour in an attempt to gain the world's water speed record lost his life in this bay Sept 29th, 1952. This memorial is erected as a tribute to the memory of a gallant gentleman by the people of Glen Urquhart. Urram do'n treun agus do'n iriosal. [Honour to the valiant and to the humble]. (Ross 2020)

After passing the John Cobb Memorial, it was on to Urquhart Castle, located on a rocky promontory near the hamlet of Strone, located south of Drumnadrochit. This historic fortification deserves commentary. It is unsure when the castle was first erected. However, in 1296, Edward I, the English monarch, invaded Scotland and his forces captured the imposing stone fortress. Urquhart Castle was later seized numerous times by various forces until the sixteenth century when Clan Grant was bestowed the "Keep." They repaired it and the Grants then constructed a five-story stone gatehouse that was blown up in

13.3 Joseph W. Zarzynski in the distance after passing the John Cobb Memorial during his 1984 ultramarathon of Loch Ness. (Credit: Pat Meaney)

1692 by government forces. Over the following centuries, Urquhart Castle fell into disrepair. In 1913, the ruins came under the cultural management of the government (Historic Environment Scotland 2019).

One of my favorite views of Urquhart Castle actually comes from the movie *The Private Life of Sherlock Holmes*, co-produced, co-written, and directed by Billy Wilder and released in 1970. Part of the 2-hour, 5-minute flick (rated PG13) was filmed on location at Loch Ness, including using Urquhart Castle (IMDb 2019). Though not a critically acclaimed movie, it nonetheless is worth watching, if just to see the amazing scenery of Loch Ness.

Running into the parking lot adjacent to A82 near Urquhart Castle was one of the stops where I met Pat to get some fluids into my body. I later noted in my journal, "At the Urquhart Castle, I felt fine…" After a quick break it was on to the heart of picturesque Drumnadrochit, just a couple of miles away. This upcoming part of the ultramarathon was also a much-welcomed downhill into the village.

As I was bounding toward Drumnadrochit, I passed Strone. At that time, this hillside community consisted of but a few residential dwellings and other struc-

13.4 Joseph W. Zarzynski running into the Urquhart Castle-parking lot during his 1984 ultramarathon. (Credit: Pat Meaney)

tures. If you were fortunate enough to have a residence at Strone, you certainly had a prime location for monster watching. Strone was ideally situated, too, overlooking the magnificent waters of Urquhart Bay, a hot spot for Nessie sightings and mid-water sonar contacts.

Two of the most enchanting residents of Loch Ness were former-RAF Wing Commander Basil Cary and his wife Winifred (aka "Freddie"). In July 1979, Tony Healy and I visited Basil and Freddie. It was the first time I met this unique couple. Three years later on August 10, 1982, Englishman Ivor "The Diver" Newby, Pat, and I paid a visit to the Carys, too. The couple were friends of many cryptozoologists—F. W. Holiday (author of the books—*The Great Orm of Loch Ness* and *The Dragon and the Disc*), Tim Dinsdale, Nicholas Witchell, Dr. Robert and Carol Rines, and others. During our 1982 visit, the Carys kept us entertained for nearly three hours telling tales about their multiple sightings of Nessie and sharing personal insight into past monster expeditions. It was one of the most pleasant and entertaining evenings that Pat and I had during our overseas travels.

Nicholas Witchell recorded in his 1974 book *The Loch Ness Story* details of one of the Nessie sightings by the Carys. On July 18, 1970, Winifred and Basil heard some people on the roadside near their Strone cottage, so they promptly investigated. Seven people were peering down the hillside toward Urquhart Bay. The party was very excited by what they were witnessing. A Nessie creature was spotted swimming on the surface of the bay by Mr. and Mrs. John Tyrrel of Nairobi, Kenya, the Carys, and several others. Winifred Cary described that incredible sighting:

By the time we got to the roadside there was just one hump visible, which looked dark brown except for where the lights from the sunset was shining on it. After a moment it turned and went away towards the far shore and [we] followed it for 15 or 20 yards perhaps and then [it] went down. But the nine of us stood and watched it, there was no question about it [at] all. The Tyrells had had a marvelous view of the head and neck and body but the head and neck had gone down by the time we got there (Witchell 1974:180-181).

Another noteworthy sighting of a Nessie animal that was observed from Strone happened on June 23, 1971. It was seen by none other than the Carys, Dr. Robert Rines, and Carol Williamson Hurley (later Carol Rines). Bob Rines recounted what they saw in his foreword of the 1972 edition of Tim Dinsdale's book—*Monster Hunt*. Rines described the 1971 sighting:

> There arose in Urquhart Bay a large hump, somewhat triangular in shape and void of any detectable fin. We measured its dimensions by telescopic comparison between this phenomenon, two-thirds of a mile away, and a nearby anchored 53-foot fishing vessel. Nessie, 20 feet long and about 4 to 6 feet out of the water at her apex, slowly moved toward us into the bay, turned, and then submerged (Dinsdale 1972:x-xi).

Running past Strone, I encountered a much-needed decline in elevation as the road moved downhill toward Drumnadrochit, a community of less than 1,000 residents. As I approached Drumnadrochit, I would soon be passing a cluster of residential houses and then I would cross the bridge over the River Enrick. In Drumnadrochit, just past Bremner's hotel, lodge, and monster museum, I darted into the parking lot that was adjacent to a small pond and stopped. Pat was waiting. I quickly toweled off and drank some water before a grueling hill, as I moved away from Drumnadrochit. My fiancée also seized the opportunity to snap a couple of photographs of that moment. My "pit stop" was on relatively flat ground. Following the water break, I began jogging along A82 on the north side of expansive Urquhart Bay. Before heading up the slope of the road, I noticed I was near Temple Pier. I was also heading toward the turnoff for Abriachan, about six miles away.

Temple Pier was nestled on Urquhart Bay that provided a protective harbor for vessels. The location afforded cover from the north winds that sometimes bashed the waterway. As I passed the dockage, I recalled Temple Pier from nine years earlier, my first experience associated with the Loch Ness monster hunting community. I was 25 years old and it was the summer of 1975. That was my first expedition to these illustrious waters and it was day one of my Highlands holiday. I had arrived at Drumnadrochit, checked into a hotel, and I was walking north along A82 when I suddenly discovered myself across from Temple Pier. I noticed an older van parked

13.5 *The author drinking water adjacent to the Loch Ness Centre & Exhibition in Drumnadrochit during his 1984 ultramarathon. The fiberglass head of a roadside-Nessie and a retired-small bathysphere are in the background behind the traffic sign. (Credit: Pat Meaney)*

on the turn-off road. The vehicle had the word Loch Ness Investigation (possibly the letters LNI), painted on the side of the cargo van.

Nine years later, in 1984, I was jogging up the same hill just outside Drumnadrochit and I thought back to that summer day in 1975. It was funny how something so mundane, a few letters painted onto the side of a van, was so eventful for me. At that moment in 1975, I realized that I was certainly in the right place and at the right time with other "kindred spirits."

Temple Pier was the footstep from *terra firma* to the water for many in the monster seeking colony. In the spring of 1968, Professor D.G. Tucker of the University of Birmingham's Department of Electronic and Electrical Engineering, with other professional colleagues, tested their sonar equipment at Loch Ness. The academic team returned a few months later, with their fieldwork operation again based from Temple Pier. The engineers directed their sonar in a narrow beam across the sweeping bay. Witchell reported in his 1974 book that their sonar pulse was transmitted at 10-second intervals. To record this, a motion-picture camera was employed to visually document the sonar results being displayed on a viewing screen. The University of Birmingham scientists reported they picked up unidentified sonar targets in Urquhart Bay, one big and fast moving. Professor Tucker did not claim the sonar results were from a Nessie. The sonar imagery must have been compelling for some of the university contingent, as they came back to the loch in 1969 and again in 1970, undertaking more acoustic fieldwork (Witchell 1974:165-167).

Not far off Temple Pier in Urquhart Bay lies a wooden shipwreck, reported by Tim Dinsdale to be in about 80 feet of water (Zarzynski 1986:5). It is a Zulu-class sailing craft, a popular Scottish fishing vessel from the late nineteenth and early twentieth centuries. I am not exactly sure when or how the shipwreck was

13.6 A 1978 Klein side scan sonar record of the Zulu shipwreck in Urquhart Bay. (Credit: Martin Klein and Garry Kozak)

first discovered, but it was sonar imaged by Marty Klein's side scan sonar while his team from Klein Associates worked with the AAS in the 1970s (Dinsdale 1984). Another Zulu-class sailing vessel reportedly is sunk off Foyers on the east side of Loch Ness (Kozak 2019).

Earlier in this chapter I wrote about Billy Wilder's 1970 movie *The Private Life of Sherlock Holmes*. One of the more-unusual aspects of this quirky film was the use of a life-size movie prop of Nessie, with those scenes initially shot on location at Loch Ness. In the movie script (Wilder and Diamond no date), the enigmatic loch was the site to test a secret submarine that was fashioned in the guise of a marine reptile from the Cretaceous period. Wilder's 1:1 scale model of Nessie was meticulously fashioned using a heavy-metal frame, leather scales, and flotation. For the movie, this behemoth model was mostly towed around Urquhart Bay by a miniature submarine, the 19-foot-long Vickers-built *Pisces II** (see footnote next page). One day in 1969, the Nessie prop unexpectedly

13.7 Filmmaker Billy Wilder's Nessie-movie model in 1969. The film prop was used in the movie The Private Life of Sherlock Holmes, *but it sank at Loch Ness during a production mishap. (Credit: Mike Macdonald, International Submarine Engineering Ltd)*

sank in Urquhart Bay. Thankfully, the *Pisces II* was either not used at the time the movie model was lost, or the tow cable was released before the fake Nessie sank. The *Pisces II* was then deployed to try to locate the sunken "monster," but the underwater search was unsuccessful. Ironically, earlier that summer before the monster-model mishap, two pilots aboard the *Pisces II*, in about 500 feet of water, reported that their underwater vehicle's sonar picked up a swimming object about 40 feet long (*Vancouver Sun* 1969). Reportedly, the monstrous mannequin cost the production company $25,000 (worth $177,000 in 2020), a considerable sum of money in 1969. Billy Wilder was forced to finish shooting the last scenes of the inanimate monster at Pinewood Studios in London (Zarzynski 1986:37-39, 55-57).

13.8 Pisces II, *a 19-foot-long submarine, was employed at Loch Ness in 1969 to tow a Nessie-movie model around Urquhart Bay during the filming of a Billy Wilder movie about Sherlock Holmes. (Credit: Mike Macdonald, International Submarine Engineering Ltd)*

---

*In 1969, there was second submarine at Loch Ness. The *Viperfish* was designed and built by Dan Taylor, a Texan. The yellow sub was deployed to find Nessie and the operation was sponsored by the LNPIB and *World Book Encyclopedia* (Zarzynski 1986:55-57).

Since 1975, I have been enthralled by this movie model. In my second book, *Monster Wrecks of Loch Ness and Lake Champlain* (M-Z Information, 1986), I had a couple of chapters devoted to the 1969 sinking of the Nessie look-alike. I also included a map that showed the approximate location where the movie-monster replica reportedly plunged to the loch bottom. I had interviewed Paul D. Herbert from Cincinnati, Ohio, who went to Loch Ness in 1985. The Ohioan met a taxi driver, one of several cab operators, contracted to transport actors and movie-production personnel around the area during the filming of Wilder's movie. Herbert's cabbie showed him the location where the movie prop was lost. I was able to put the proverbial "X" on the map of Loch Ness that is in my 1986 book. That showed the spot where I believed the inanimate Nessie sank (Zarzynski 1986:37-39).

Years later, in 2012, I collaborated on a proposed documentary project with a friend, Peter Pepe of Pepe Productions (Glens Falls, New York), an award-winning documentarian. We had produced three documentaries on shipwrecks and underwater archaeology. Peter and I wanted to do a television documentary on Billy Wilder's Nessie "wreck." Our working title was "Sherlock's Missing Loch Ness Movie Monster." Here is a section from our proposal for acquiring television sponsorship for the documentary:

> The Pepe Productions' television proposal will seek funding for both the production of the show as well as using remote sensing and sonar expert Vincent J. Capone (Black Laser Learning) as the technical expert to help coordinate the search for this…movie prop that sank in Loch Ness. Due to the waterway's vast depths…the team proposes using an autonomous underwater vehicle (AUV)* equipped with side scan sonar and video capability to locate and visually document the "Nessie" movie monster prop. This type of specialized equipment will be more suitable than deploying traditional side scan sonar towed from a craft-of-opportunity to locate the movie prop and then use a tethered underwater video robot,

---

*An autonomous underwater vehicle (AUV) is a drone programmed to drive underwater, collecting data. The AUV is directed without real-time human control (Zarzynski 2019:203).

known as a remotely operated vehicle (ROV), to visually document the inanimate "monster" (Pepe and Zarzynski 2012).

Unfortunately, Peter Pepe and I were unsuccessful in getting the documentary proposal greenlit* by a television channel. This was one of the biggest disappointments in my documentary career. Pepe Productions was a superb documentary filmmaking company. Including a documentary on colonial-era cannons that we completed in 2016, I collaborated with Peter and his son Joe on four documentaries. Thus, I knew firsthand, Pepe Productions' vast expertise in their field. So, I believed if we had secured funding, we would have done an admirable job on that proposed television production. However, though we did not get our 2012 Loch Ness documentary proposal greenlit, Left Coast Press of Walnut Creek, California, did publish our book—*Documentary Filmmaking for Archaeologists*. So, Peter Pepe and I did have some documentary success in 2012.

Then four years later in 2016, the Nessie movie double, approximately 30 feet long, was finally discovered on the deepwater bottomlands of Loch Ness. The find was made during a joint sonar survey undertaken by Adrian Shine's Loch Ness Project and Kongsberg Maritime. The team employed a Kongsberg Maritime autonomous underwater vehicle (AUV) equipped with sonar to locate the missing film monster (Victor 2016).

The incredible discovery of Wilder's movie prop in 2016 got me wondering where the inanimate creature had been found in the loch. In 2005, Garry Kozak, employed by Klein Associates, used a more sophisticated Klein side scan sonar then utilized during the 1970s. The 2005 survey deployed a towed side scan sonar, rather than Adrian Shine's 2016 fieldwork that had the luxury of an AUV. In 2005, Kozak's sonar scanned the full length of Loch Ness. With Marty Klein and others aboard the research boat, the 2005 group ran the Klein sonar at a 100 kHz frequency range. Ironically, the searchers in the year 2005 also included Shine. The workboat that year conducted a single pass down the middle of the 22 ½-mile-long

---

*The word "greenlit" is terminology frequently bantered about in the movie and documentary film industry. If a movie or documentary gets formal permission and funding to move ahead with production, it is said the project is greenlit.

waterway. Amazingly, the 100 kHz frequency range was pretty much able to sonar image the full width of the loch. However, that was without acquiring the resolution details that a 500 kHz side scan sonar would have accomplished. A 500 kHz side scan sonar gives higher resolution and with more details, but at the expense of not having as great a range as a 100 kHz sonar unit. A 500 kHz side scan sonar operation to scan all of Loch Ness would have required a survey of many days with numerous sonar passes undertaken.

Following the discovery of the submerged film model in 2016 by Loch Ness Project/Kongsberg Maritime, I asked Garry Kozak to refer to the map published in my 1986 book—*Monster Wrecks of Loch Ness and Lake Champlain*. The annotated map of Loch Ness gave my best estimate, based on Paul D. Herbert's letter, as to where the movie model sank in 1969. After reviewing his sonar records from 2005, Kozak informed me that my location was in the very vicinity of the sunken movie prop. The side scan sonar fieldwork in 2005 had actually recorded the target, but the lost Nessie model was not recognized by the sonar team at that time. That was probably because the sonar unit was operating at a long-range frequency making it difficult to interpret smaller sonar details. The

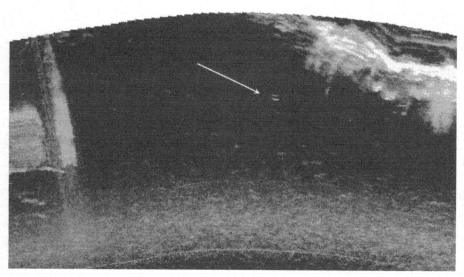

13.9 A 2005 Klein side scan sonar record that shows the Billy Wilder movie monster that sank in Loch Ness in 1969. (Credit: Garry Kozak)

movie monster target was there, but it was simply too little to perceive what it actually was. Nonetheless, Adrian Shine's sonar expedition in 2016 deserves full credit for discovering Billy Wilder's Nessie double. Shine's team, the Loch Ness Project/Kongsberg Maritime, had accomplished a significant-sonar feat.

After running past Temple Pier on August 22, 1984, and striding up the crest along A82 just north of Drumnadrochit, I peered off to my right across the wide bay. What an unbelievable sight. There across the cove stood the ancient stone ruins of Urquhart Castle with its noble view in its totality. I had previously examined the castle from many angles and perspectives, but this brief glance was special. In my journal entry that night, penned several hours after the ultramarathon, I wrote: "...from Drumnadrochit to the lay by north of that bay I felt fatigue."

As I was plowing ahead, stride-after-stride, I wondered if I would soon be "hitting the wall." That's the time in a marathon or ultramarathon when weariness begins to batter the psyche and the physical abilities of a long-distance runner. "Hitting the wall" is probably the greatest fear of the average marathoner or ultramarathoner. It is toward the last few miles of the distance race when a loss of energy suddenly descends upon the runner. It is caused by a variety of factors. Most often it is due to the depletion of glycogen in your muscles and liver. That is why proper training for your run, carbohydrates-loading the day before your race, and fluid intake before and during the athletic event are so critical. During an ultrarun, fluid intake should likewise include carbohydrates-enriched drinks and sometimes even mixed with a light snack for fuel. Fatigue is one thing, but "hitting the wall" is an overwhelming sensation that frequently has caused novice and experienced distance runners to literally stop participating in a race. It takes a lot of determination and will power for marathoners and ultramarathoners to battle through this running phenomenon.

I estimated that I was nearing about 20 miles or so, that is, over two-thirds of the way into the ultrarun. I had over eight miles to go to reach my finish line, a spot well beyond the village of Lochend. I knew I would have to increase my fluid intake and like other distance runners, dig deep into my previous experience. Since I hadn't gotten in a couple of 20- to 22-mile runs, I wondered if I would pay the price of not enough long-distance training jogs for this ultramarathon?

Chapter 13: My One-Person Loch Ness Ultramarathon—
John Cobb Memorial to Beyond Urquhart Bay

At the lay by off route A82 that is near the northern edge of Urquhart Bay, I received water and other fluids from Pat. I had also stopped about a mile or so earlier at the eastern edge of the Drumnadrochit Hotel property, where a full-scale Nessie model was positioned, floating in a tiny pond. My fiancée was doing a praiseworthy job driving to the lay by pull offs ahead of me, waiting to offer me water, and if time permitted, to capture a photograph. Pat was certainly doing her part. It was now up to me to finish the last miles of the ultramarathon. It might sound trite, but those unique opportunities to gaze over Loch Ness and think about Nessie were beneficial. It took my mind off the growing fatigue and pain that I was feeling.

Chapter 14

My One-Person Loch Ness Ultramarathon—
the Clansman Hotel to Across from Dores

Wednesday—August 22, 1984 (continued)

It was now on to the popular Clansman Hotel, another well-known tourist lodging adjacent to the shores of Loch Ness. The Clansman Hotel in August 1984 was a popular motor lodge located north of Drumnadrochit. The inn also had a restaurant with broad picture windows that made it a desired eatery for visitors and locals. In September 1985, the year after my ultrarun, the Clansman Hotel served as the media center for the recovery project of N2980, the Wellington bomber lost in Loch Ness.

That warplane, skillfully piloted by Paul Harris, had participated in the historic December 18, 1939 air raid known as the Battle of Heligoland Bight. The daylight-RAF operation upon German targets by a squadron of 24 Wellingtons met with disaster. Ten planes were shot down, two ditched into the sea, and two returned early (The Loch Ness Wellington Association Ltd. 1985). As a result, the RAF then went only to nighttime raids on the Germans, which the British leadership believed would not be so deadly to their air corps (Holmes 1991:3).

On New Year's Eve, 1940, the plane was on a training mission from Lossiemouth, an air base in northeastern Scotland. The bomber unexpectedly lost power in one of its two engines. Sorrowfully, one of the six airmen who had been ordered to parachute out of the airplane when the mechanical problems first occurred, died of his injuries. The pilot and another aviator successfully ditched the

14.1 A World War II British Wellington bomber and flight crew before a 1939 raid on Germany. (Credit: Imperial War Museum—London)

warplane onto the surface of Loch Ness. The two-RAF personnel aboard the plane were able to launch a small dingy from the floating aircraft and they paddled to shore. Soon after, the twin-engine bomber sank (Holmes 1991:93-94).

That event was all but forgotten until one day in 1976, when Martin Klein and Charles Finkelstein (Klein Associates, Inc.), working with the Academy of Applied Science, used a side scan sonar to discover the plane sitting upright in deepwater. Klein and Finkelstein were not looking for the British bomber. Finding the airplane was a notable by-product of their Nessie fieldwork.

The sunken plane was raised in September 1985. I traveled to Scotland for that recovery operation and thus, I spent a lot of time in the Clansman Hotel, where the recovery team met with the media. In front of the Clansman Hotel was a modest marina. Later in the mid-1990s, the boat berth complex was the site for a submarine-tour enterprise that could take five paying tourists at a time on voyages into the depths of the mysterious loch. This was "cryptotourism" at its optimum. A voyage in the 32-foot deep-diving submarine reportedly cost about $102 and the underwater excursion lasted for an hour (*New York Times* 1994).

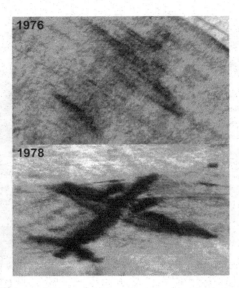

14.2 Two Klein side scan sonar records of a Wellington bomber sunk in Loch Ness. One from 1976 and the other with improved sonar technology from 1978. The plane ditched onto the loch surface during a training mission in 1940, after losing power in one of its two engines. (Credit: Martin Klein and Garry Kozak)

Years later I would sometimes tell myself that I wished I had that opportunity to explore Loch Ness's semi-opaque waters from the portal of a submarine or a submersible.*

As I passed the Clansman Hotel on my run, I had about four miles left in the ultramarathon. I was obviously tired, but surprisingly, I was feeling better than a few miles back. Maybe it was the flatter ground or just an adrenaline burst knowing that I had finished over 85 percent of my solo run. I soon would be approaching the turn off A82, toward Abriachan.

The area around Abriachan is rather famous in the annals of Nessie. That is because of a reported land sighting of one of the monsters in 1934, observed by a 21-year-old veterinary student. W. Arthur Grant lived in Glen Urquhart, the narrow valley west of Drumnadrochit. It was January 5, 1934, and at 12:30 a.m. Grant was riding along route A82 aboard a motorcycle. He was returning home from Inverness. Grant, who declared he was completely sober at the time, was driving south near Abriachan; others believe he may have been closer to the pres-

---

*A submarine is an underwater craft that requires no surface vessel for support. A submersible, however, is aided by another watercraft, a shore team, or a type of platform. Often a submersible is tethered by a cable to a surface-support vessel.

ent-day Clansman Hotel. There was an illuminating moon that night that lit up the roadway ahead of him. Suddenly, Grant came to an abrupt stop when he spotted of all things, a formidable dark object 40 yards ahead in the bushes. Grant later described what he saw:

> I was almost on it when it turned what I thought was a small head on a long neck in my direction. The creature apparently took fright and made two great bounds across the road and then went faster down to the loch, which it entered with a huge splash. I jumped off my cycle and followed it but from the disturbance on the surface it had evidently made away before I reached the shore. I had a splendid view of the object. In fact, I almost struck it with my motorcycle. The body was very hefty. I distinctly saw two front flippers and there seemed to be two other flippers which were behind and which it used to spring from. The tail would be from five to six feet long and very powerful; the curious thing about it was the end was rounded off — it did not come to a point. The total length of the animal would be 15 to 20 feet (Witchell 1974:137).

14.3 This drawing interprets W. Arthur Grant's sighting of a Nessie crossing route A82 at Loch Ness in January 1934. (Credit: August Johnson)

This sighting of a possible Nessie always intrigued me. If not a hoax, Grant's sighting could have been of a known animal that in the darkness, was misconstrued to be the Loch Ness monster. More recently, Tony Harmworth believes the Grant land sighting of a Nessie may have been a fabrication perpetrated by W. Arthur Grant (Harmsworth 2012:94-95).

However, Grant, reported to be a veterinary student at the University of Edinburgh, presumably knew fauna. Also, he told a newspaper reporter that he observed the animal completely out of water and gave an estimate of its total length at "15 to 20 feet." I always felt that if Nessie and Champ are real animals, that many eyewitnesses might be giving superstar-size status to these lake monsters. The length of 25 to 35 feet often reported by eyewitnesses may be too large. If these enigmatic creatures exist, a length of 17 to 20 feet seems more realistic, which matched Grant's sighting. Furthermore, a plesiosaurus did not reach the length of a 50-foot plesiosaur. Rather, a plesiosaurus, an early type of plesiosaur, had a length of only about 11 to 15 feet.

As I passed the Abriachan exit off route A82, I contemplated Grant's land sighting of a monster. At that time I thought, "Heck, I really am running where Nessie may have once tread, 50 years earlier. How cool was that!"

Half a mile further on A82 and I was nearly across the waterway from the village of Dores, on the eastern side of Loch Ness. After my ultrarun that day I wrote in my journal: "At Dores I felt very excited that I could break 4 hours and 30 minutes" for my run.

Chapter 15

My One-Person Loch Ness Ultramarathon—
the Final Stretch to Past Lochend

Wednesday—August 22, 1984 (continued)

Dores is a bedroom community for some of the people working in the city of Inverness. There is a sandy beach at Dores, too. This village has a hotel with a restaurant and pub, appropriately named "The Dores Inn." The lodging and eatery are tucked up to the shore of the northeastern part of Loch Ness.

It was at Dores in late 1933, spurred by the news frenzy of the Mackays' April sighting of a sizable water creature in the loch, that a farcical hoax was created. Marmaduke Wetherell, a big-game hunter and a Fellow in the Royal Geographical Society in London, along with his photographer Gustave Pauli, were hired by the *Royal Mail* newspaper of London. Their photo-journalism assignment was to travel to Loch Ness to solve the monster mystery. Shortly after arriving at Loch Ness, Wetherell claimed he discovered and photographed a strange footprint along the shoreline at Dores. He adamantly asserted it was from a Nessie animal that had crawled up onto shore.

On December 21, 1933, Wetherell was quoted in the *Daily Mail* newspaper:

"It is a four fingered beast and it has feet or pads about eight inches across. I should judge it to be a very powerful soft-footed animal about 20 feet long" (Witchell 1974:60).

The chicanery reportedly involved the use of a stuffed hippopotamus foot, possibly from an unorthodox umbrella stand, that was imaginatively employed to fashion the Nessie footprint. It was not completely clear what role, if any, the megafauna tracker played in this fantastic stunt (Tikkanen 2020).

However, the Wetherell saga reportedly did not end with this footprint hoax. For decades, the most famous image of Nessie was the 1934 "surgeon's photo" (aka Wilson photo), supposedly snapped by Robert Kenneth Wilson, an English physician. Over the years, this photograph has been published in numerous newspapers, magazines, journals, and even in my 1984 book, *Champ—Beyond the Legend*. In 1994, it was reported that Wilson's photograph was a prank conceived by none other than revenge-seeking Marmaduke Wetherell and others. Reportedly, decades after the 1934 photograph, one of those involved in this purported deception, provided details in a death-bed confession. That person declared the photograph was actually a small plastic and wooden head/neck of a plesiosaur-like animal affixed to a toy submarine (Tikkanen 2020). Even with the 1994 revelation, there are some who dismiss the hoax

15.1 The controversial-1934 "surgeon's photo" of Nessie. Is this famous photograph of a Nessie animal or was it an elaborate hoax? (Credit: Robert Kenneth Wilson)

assertion for the "surgeon's photo." One person who believed this photograph may be that of an actual Nessie creature is cryptozoologist Richard Smith. He cited the lack of a toy submarine, technically sophisticated from that era, that would have been capable of being the platform for such a photographic scam (Lyons 2000).

While running across from Dores in 1984, I also thought about another tantalizing enigma of Loch Ness, clusters of perplexing stone circles lying mostly in shallow water at the northeast end of the waterway. One of the first articles I read about these baffling inanimate configurations was the Martin Klein and Charles Finkelstein report entitled "Sonar Serendipity in Loch Ness" (Klein and Finkelstein 1976:45-57). Their article appeared in the December 1976 issue of *Technology Review* magazine, a respected MIT publication.

In 1976, Klein's team recorded these "mysterious circles" during sonar mapping of parts of Loch Ness. Divers donned scuba gear and inspected one or more of

15.2 Charlie Finkelstein and Marty Klein examine their sonar records of mysterious stone circles found on the bottomlands of Loch Ness in 1976. (Credit: Martin Klein)

these unusual sonar targets. The frogmen confirmed one anomaly was a circle-like stone configuration of about "10 meters," or 33 feet, in width. After the discovery of these apparent stone circles and rock piles, the sonar technicians began to refer to them as "Kleinhenge." A couple of these stone circles were dubbed "Kleinhenge I & II." That was a play-on-words to the Klein side scan sonar that discovered the underwater stone circles and to Stonehenge.* (Klein and Finkelstein 1976:49, 52-53). Some have suggested these were not paleolithic configurations, but were created by dredge spoils dumped from barges, possibly from the construction of the Caledonian Canal or some other major-work project during the nineteenth or twentieth centuries (Zarzynski 1986:81-82).

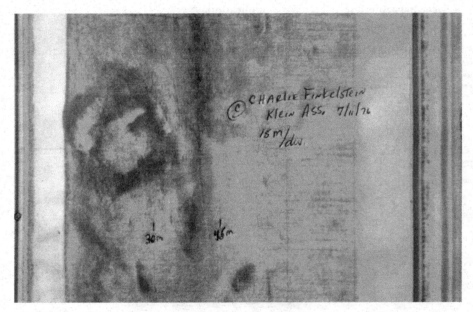

15.3 A 1976 Klein side scan sonar record, at left, of one of the stone circles on the bottom of Loch Ness. (Credit: Martin Klein/Charles Finkelstein)

---

*Stonehenge is one of the most famous prehistoric lithic monuments in the world, located near Amesbury, England, about 75 miles from London. The ring of tall-standing rocks is believed to have been erected over stages about 4,500 to 5,000 years ago (English Heritage 2019). In 1975, I had the pleasure of visiting Stonehenge, 11 years before it was listed as one of the first UNESCO (United Nations Educational, Scientific and Cultural Organization) World Heritage Sites.

J. D. Mills of Underwater Instrumentation, a British firm that specialized in commercial diving and sonar equipment, worked for the AAS examining these underwater stone formations. In September 1985, Mills described one of the rock assemblages on the loch floor as being about "30 ft. across and roughly 3-4 ft. high." Furthermore, Mills noted, "The general shape is that of half a doughnut split at right angles to its axis and resting on the bottom sediment" (Mills 1985). Mills then reported: "All the superficial fine material seems to have been washed away by the current which must be strong in the area when the water is high... [Thus] the circular formation [is] due to the way the material slumps out of the barge onto the bottom 30 to 40 feet below... and would account for the apparently random distribution of the circles and the way in which they overlap" (Zarzynski 1986:81-82).

In 1985, P. H. Milne was with the Department of Civil Engineering at the University of Strathclyde in Glasgow, Scotland. Milne provided me with a report he authored on this subject, entitled—"Underwater Stone Circles at Loch Ness" (Milne n.d.). After his perusal of the data collected by the AAS and his own investigation, Milne concluded it was unlikely the stone formations were cre-

15.4 An underwater photograph through the peat-stained depths of Loch Ness that shows one of the stone circles lying on the bottom of the waterway. (Credit: Academy of Applied Science and Underwater Instrumentation)

15.5 Garry Kozak, an experienced scuba diver and sonar specialist, prepares to dive at Loch Ness in the 1970s to inspect some submerged stone circles. (Credit: Garry Kozak)

ated by ancient-human hands since the general depth of the lithic oddities was 10 meters (33 feet). Thus, he said it was doubtful the loch-bed area had once been dry land. Moreover, due to their depth, it was improbable these "could be the bases of crannogs [artificial islands]" (Milne n.d.). Another possibility Milne offered to explain the submerged stone formations was "that the polygonal patterns could have been formed by periglacial ice wedges during the retreat of the last ice age and before the inundation of Loch Ness" (Milne n.d.). The University of Strathclyde professor recommended "further archaeological work" to determine their exact origins (Milne n.d.).

Navigation on Loch Ness via the Caledonian Canal began in 1818. In 1821, to deepen the channel between Loch Dochfour and the north end of Loch Ness, a "dredging machine" was employed (*The Caledonian Mercury* 1821). However, few details were reported in newspapers at that time on what was done with the clay, stones, and tree trunks that were removed.

Marty Klein has not been so quick to conclude that the "stone circles" that his team recorded with sonar back in the mid-1970s were all due to dredge spoils. In 2020, Klein added: "There is another [sonar] image [we made] with a large circle and a row of circles and a circle of circles that I called Kleinhenge III. It tends to be forgotten. Not sure how this can be dredge spoil" (Klein 2020).

It should be added, other curious side scan sonographs, appearing to show inanimate objects, likewise have been recorded at Loch Ness. These include long linear rows of unidentified features found on the loch bottom. One of these stretches for a considerable distance, mostly near the centerline of the loch (Kozak 2005).

According to Garry Kozak, who was the side scan sonar operator on a 2005 remote sensing project at Loch Ness, this row appears as a long line of dots on the sonar records. The objects on the loch bed are "between 10–14 meters" [32.8–45.9 feet] in diameter and "are spaced an average of 70 meters [229.7 feet] apart" (Kozak 2020).

Were they formed by dredge spoils from the construction of the Caledonian Canal, nearly two centuries earlier? Are they refuse deposits from decades ago, possibly from the aluminum factory on the shore of the waterway? Could they be related to some secretive monitoring system from the Second World War? In 2020, Marty Klein suggested that these unusual sonar features are most deserving of future examination, study, and analysis.

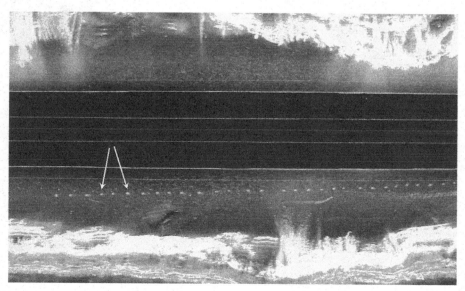

15.6 2005 Klein side scan sonar record of a line of mysterious-sonar contacts lying on the bottom of Loch Ness. (Credit: Garry Kozak)

When I was but a few miles from the finish line of my ultrarun, I thought a lot about the famed AAS expedition at Loch Ness in 1976. The anticipation that a full-scale search operation might be meaningful, persuaded the newspaper hierarchy of the prestigious *New York Times* to co-sponsor the Nessie-hunt endeavor. Moreover, that summer, *National Geographic Magazine* sent a team of scuba divers directed by scientist Dr. Robert D. Ballard (Woods Hole Oceanographic Institution) to conduct Nessie-related fieldwork; the June 1977 issue of the publication had an article on Loch Ness and Nessie (*National Geographic Magazine* 1977:758-779). The AAS fieldwork at Loch Ness in 1976 likewise attracted other notable personalities.

One of those celebrities was Garry Trudeau, a Pulitzer Prize award-winning American cartoonist, known for his immensely-entertaining "Doonesbury" comic strip. Trudeau spent time at Loch Ness in 1976, embedded with some of the scientists. He generated several humorous "Doonesbury" comics about the much-publicized Nessie expedition that summer (Trudeau 1976).

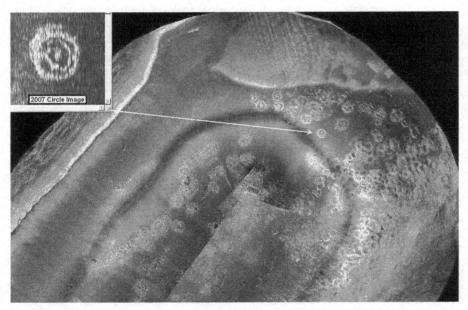

15.7 *This 2007 Klein side scan sonar record shows some of the stone circles on the bottomlands of Loch Ness in the north of the waterway. The insert (upper left hand corner) pictures a close-up sonar record of one of the underwater features. (Credit: Garry Kozak)*

In 1977, Dennis L. Meredith, the Managing Editor of *Technology Review* magazine, wrote a book about the AAS field project in 1976 at Loch Ness. Meredith's work was entitled *Search at Loch Ness: The Expedition of the* New York Times *and Academy of Applied Science.* The publication provided further insight into the sonar surveying and other field operations at the Scottish waterway during the USA's bicentennial year. Meredith gave high praise to the group's sound-imaging technology, declaring, "Sonar proved to be the star of the 1976 expedition" (Meredith 1977:134).

The underwater efforts that year consisted not only of Klein side scan sonar to survey the loch looking for a carcass of a Nessie, but also sonar scanning of the loch-bed geology. For this, Klein Associates brought to Scotland the company's new "combined side scan sonar/sub-bottom profiler system" (Klein and Finkelstein 1976:47). The sonar unit operated at 100 kilohertz (kHz) frequency, using a 100-meter (328 feet) light-weight cable, this for surveying wide swaths of the loch bottom. Moreover, the Klein sub-bottom profiler sonar mapped sections of the loch floor with some penetration of the bottom sediment that was directly beneath the towfish. The sub-bottom profiler worked at 3.5 kHz with a conical beam of 50 degrees (Klein and Finkelstein 1976:47). The sonar surveying was conducted off Lochend and Dores Bay at the north end of the loch, in Borlum Bay off Fort Augustus, in Urquhart Bay adjacent to Drumnadrochit, and in the waters around Cherry Island at the south end of the loch (Klein and Finkelstein 1976:47).

15.8 Marty Klein in the 1970s with his side scan sonar/sub-bottom profiler at Loch Ness. (Credit: Martin Klein)

According to Klein and Finkelstein, several intriguing targets were noted during their sonar fieldwork. One of those remarkable sonar targets, mentioned earlier in the book, was the "The Average Plesiosaur."

"The Average Plesiosaur" sonograph was not happenstance. In the 1970s, prior to the Klein side scan sonar surveys at Loch Ness, Marty Klein wondered "whether there might be bones of dead 'monsters' in the loch" (Klein 2019). Therefore, Dr. Christopher McGowan, a paleontologist at the Royal Ontario Museum in Toronto, brought some mammoth bones that were sunk in a New England pond for sonar tests. Sonar technicians deployed a Boston Whaler watercraft equipped with a Klein side scan sonar. The sophisticated apparatus successfully recorded "some traces of the bones" (Klein 2020). Furthermore, in a Maine lake, the sonar sleuths simulated an underwater-swimming Nessie. They towed an old, stuffed moose head mounted on a wooden base through the water as they practiced acquiring its sonar signature. Thus, Klein and his colleagues had done their pre-Loch Ness survey "homework." They were fully prepared to sonar search at Loch Ness, hoping to locate the remains of a deceased Nessie lying on the waterway's bottomlands or to detect a live one swimming in the water column (Klein 2020).

Complementary sonar fieldwork was likewise undertaken at Loch Ness in 1976, too. That was mainly under the direction of "Doc" Harold Edgerton, also a member of the AAS. "Doc," as he was affectionately known, had been an electrical-engineering professor at MIT and was Marty Klein's mentor.

Just a few miles to go in my 1984 run and I was still moving at a decent pace for an "average" 34-year-old ultramarathoner. The finish line for my ultrarun of Loch Ness was a spot past Lochend. As I crossed my designated finish line, I hit the stop button on my Casio wristwatch to record the time for my ultramarathon.

From start-to-finish, my clock time (gun time) was 4 hours, 23 minutes, and 26 seconds (4:23:26). We had earlier clocked the distance using the odometer on the rental vehicle. The length, including my running into the lay bys and off other spots along the A82 roadway, was 28.5 miles. That was about 2.3 miles farther than the standard marathon. I may have been the first person to run the full length of Loch Ness from a starting spot in Fort Augustus, that was south of the head of the loch, north to a location that was well past the end of Loch Ness.

Chapter 16

Celebrating My Loch Ness Ultrarun

· ·

Wednesday—August 22, 1984 (continued)

At the conclusion of my one-person run at almost 11:30 a.m., Pat and I went back to a road sign that read—Lochend. Exhausted, but nevertheless silently elated with the outcome, I posed next to the road marker as Pat snapped my photograph. She didn't take many photographs of me running that day as her priority was to drive ahead to pass out water as I came along. Still, that photographic image was quite the symbolic visual, standing at a road sign marking the settlement of Lochend. I was also pleased that my one-person ultrarun raised a little money for a British charity, the National Society for the Prevention of Cruelty to Children, a

16.1 A tired Joseph W. Zarzynski after completing his solo ultrarun, 28.5 miles, along route A82 at Loch Ness in 1984. (Credit: Pat Meaney)

119

charitable organization suggested to me by Tim Dinsdale. I did not collect as much money as I had hoped, only a few donations from some friends and myself. I lacked the skills of being a competent fundraiser, but though a paltry sum, the donations were for a great cause.

I wondered why Tim Dinsdale recommended that charitable organization. Tim never said why. I think it could have been because as a child aboard a packet ship traveling in the Pacific Ocean, the vessel was hijacked by Chinese pirates. It is an amazing story and it deserves to be shared.

Tim Dinsdale was born in Wales in 1924, and as a child, his parents took the family to China where his mother and father were missionaries. The "Middle Kingdom," as China was referred to years ago, was a turbulent land in the 1930s, besieged by civil war and lawlessness. In 1935, Tim's parents sent their daughter Felicity as well as brothers Peter and Tim by steamer from Shanghai, on a voyage along the Chinese coast. Their destination was the city of Chefoo, today known as Yantai. The oceanic cruise was about 500 miles. The vessel, the SS *Tunghow*, carried not only Chinese travelers, but also dozens of foreign children returning to China for their spring-term studies after a seven-week vacation.

Stowaway bandits aboard the steamer seized the packet ship on the first day of the journey. The pirates grabbed $250,000 in one-yuan notes that were being transported from London to China. However, the brigands were unaware that the paper currency was only partially printed. The final printing was to be completed by the Bank of China. When word got out that pirates had commandeered the non-military craft, the British Royal Navy ordered warships to seek out the SS *Tunghow*. This massive search-and-rescue operation also included a British aircraft carrier, the HMS *Hermes*, laid down in 1918, and commissioned six years later. When planes from the *Hermes* finally spotted the hijacked steamer, they flew over the *Tunghow*. Seeing that formidable display of British aerial firepower, most of the thieves abandoned the captured steamship. Shortly after that, the remaining buccaneers retreated to lifeboats and headed to shore. The hostages were rescued and the Dinsdale children were reunited with their parents (A. Dinsdale 2013:20-26). This traumatic event must have left a lasting impression upon Tim. It is possible that this 1935 episode was Tim's reason for backing a British charity that supported children's welfare.

16.2 Cryptozoologists Tim Dinsdale (right) and Joseph W. Zarzynski (left) along the shore of Loch Ness in 1979. (Credit: Tony Healy)

Additionally, prior to the ultrarun, I designated Tim Dinsdale as my "honorary coach" in absentia. Not only did he pour countless hours into scanning the waters of Loch Ness looking for Nessie, he also was a prolific author on the subject, too. Dinsdale's books: *Loch Ness Monster* (1961), *The Leviathans* (1966), *Monster Hunt* (1972), *The Story of the Loch Ness Monster* (1973), and *Project Water Horse: The True Story of the Monster Quest at Loch Ness* (1975).

A few hours after my ultra, I wrote in my journal: "I estimate from counting cars that approx. 350 vehicles passed me [heading south and in my lane]. Of those, I probably stopped or at least slowed down my pace on 275 of them." My time of 4 hours, 23 minutes, and 26 seconds (4:23:26) calculated to a pace of 9 minutes, 15 seconds per mile. That included all the numerous times I had to pause my running and leap over the guardrails along the loch-side edge of the road. My gun time also encompassed all my water stops and my three calls from "Mother Nature." I was overjoyed that I never stopped to walk. I really was one lucky athlete not to have tripped, fallen, gotten injured, or even worse, killed by a collision with a car, lorry, or bus. Thank you, guardian angel.

After my distance trot, I toweled the perspiration off my legs, arms, and torso, washed the sweat from my face with water, changed into some "civvies," put on deodorant, and I continued to drink water to replenish fluids lost during the run. Pat and I then drove down to the Clansman Hotel. We were given a table at one of the windows that overlooked the waterway for our well-earned lunch. I can't remember my meal, but if I could turn the clock back, I would bet it was absolutely delicious. Pat, too, seemed pleased with her driving. She certainly did A+ work that summer morning.

After lunch, but before we returned to our room at the Benleva Hotel, we stopped at the Drumnadrochit Hotel to give Tony Harmsworth our good news. Unfortunately for us, Tony was with the Loch Ness Project. Therefore, I wrote a note informing him of the successful completion of my run, including my time for the one-person race. Harmsworth later put the ultramarathon into perspective. He noted that I "may be the first person to complete this run along route A82" (Zarzynski, 1984d:1).

Then back to normalcy. After I showered, and rested for awhile at the inn, we entered the world of the mundane. We spent the latter half of the afternoon washing our dirty clothes at a nearby laundromat, especially my sweaty running garments from that morning. After dealing with grubby laundry, Pat and I dashed off for another celebratory event, a fine dinner at the Loch Ness Lodge Hotel in Drumnadrochit. We both slept quite well that evening, knowing we had accomplished one of our major Loch Ness goals for our 1984 trip. It was now time to move away from ultrarunning and back to the field of cryptozoology.

Chapter 17

1984—Post Ultrarun in Scotland

· ·

Thursday—August 23, 1984:

With the big run out of the way, we then focused solely on monster hunting. Pat and I got up the next day at our normal time. Understandably, I felt lethargic and my legs were sore from the 28.5-mile run. However, this was no different than how I felt after my other marathons and longer races. I knew it would take three to four days, maybe more, to recover. We had breakfast at the Benleva Hotel and then we went into the heart of the village of Drumnadrochit. We mailed some postcards,* cashed traveler checks at a bank, and then went to the Loch Ness Centre & Exhibition.

At the museum and hotel complex, Pat and I chatted with Tony Harmsworth and Ronnie Bremner during coffee at the facility's restaurant. I later wrote in my journal: "Ronnie is rather interested in seeing a Loch Ness run (marathon or ultramarathon)." I was not surprised by Bremner's comment since he enjoyed all types of sports. Little did I know at that time, but Ronnie Bremner's words were quite prophetic.

Years later, after the death of Ronnie Bremner in 2001, Loren Coleman, a New England cryptozoologist, wrote an informative obituary about Bremner (1941-2001). Loren had a keen penchant for drafting insightful obituaries for those in

---

*This was the era before emails, cell phones, and texting. Since overseas postage for a letter from Britain to the USA was expensive, it was often cheaper to buy postcards and mail them postcard rate. Back then, people enjoyed receiving postcards, especially from faraway Loch Ness.

17.1 Ronnie Bremner, a Drumnadrochit, Scotland entrepreneur. His Loch Ness businesses expanded Nessie-related tourism. (Credit: Martin Klein)

the cryptozoological community. Coleman published many of these obituaries in his blog—*CryptoZooNews*. Loren's death notice about Ronnie Bremner included some of the following information.

Ronald Bremner was born during World War II, the son of hoteliers. Loren described Ronnie as a "a topnotch sportsman, playing rugby, tennis, table tennis and golf to a high standard." As already noted, Bremner collaborated with Tony Harmsworth to open the highly successful Loch Ness Centre & Exhibition, a facility that not only interpreted the Loch Ness monster phenomenon, it also did a stellar job explaining the loch's limnology (study of inland waters). Later in life,

Ronnie became afflicted with cancer and turned the management of his businesses over to his family (Coleman 2003).

Ronnie Bremner was a shrewd entrepreneur and a visionary, too, for understanding the tourist market for the Scottish Highlands. A person generally "ahead of the curve," he sadly would not live to witness his 1984 suggestion of one day seeing a marathon race at Loch Ness. Ronnie Bremner passed away on December 1, 2001, ironically just several months before the inaugural Loch Ness Marathon in 2002.

Following our August 23, 1984 chat with Harmsworth and Bremner, Pat and I departed the charming village of Drumnadrochit. We headed our automobile south along A82 toward Fort Augustus. I later recorded in my journal: "All along A82 we drove, casually with an eye on the loch…hoping to get that final look… and possibly at Nessie." The rest of our 1984 trip to Scotland would be dominated by an excursion to Loch Shiel, hoping to gain greater insight into this alleged home of cryptids called "Seileag" (Zarzynski 1984b:50-54).

Chapter 18

Loch Shiel and Seileag

. . . .

Thursday—August 23, 1984 (continued)

Our expedition to Loch Shiel, following my ultramarathon, deserves coverage as it further puts into perspective the life of a cryptozoologist. Obviously, lake monster hunters are a unique and peculiar breed. If we can't find a cryptid in one place, we are apt to look for their scent at another reported monster site. I wanted to know if Loch Shiel, located in a remote section of western Scotland, could have similar unidentified animals like Nessie? Our drive to Glenfinnan, a hamlet at the east end of Loch Shiel, was along route A82 before we turned onto route A830 at Fort William. Our trip to Glenfinnan was a little over 60 miles.

Glenfinnan is a special place for Scots. It is where Charles Edward Stuart ("Bonnie Prince Charlie") raised his standard in 1745, known as the "Jacobite rising of 1745." The revolutionary hoped to gain the British throne. Several months later in 1746, "the Young Pretender," as he is known, and his supporters were defeated by a British government army at Culloden, Scotland (National Trust for Scotland 2020).

On our trip to Glenfinnan, we motored near Loch Lochy, a waterway of the Great Glen and a reputed monster dwelling, too. We pulled our automobile into a lay by to peruse this beautiful loch. There we spent time doing shore watching, looking for Lizzie, the maybe-monsters of Loch Lochy. The waterway is long and narrow, about 10 miles in length and no more than a mile wide. Loch Lochy has a maximum depth of about 531 feet (Zarzynski 1982:79). Like several other Scottish waters, it has folklore of having cryptozoological unknowns like at Loch Ness, Loch Morar, and Loch Shiel.

Looking over the waters of Loch Lochy, I recalled a report from the mid-1970s of a Lizzie sighting. After getting back home from our Scotland trip, I reviewed my files on that, a sighting by a husband, wife, and their child. The trio was from nearby Fort William. Their encounter was on September 30, 1975, at about 1:20 p.m. The loch was "very calm and flat." They first noticed an "unusual wake in the water." Then they observed an animate object about "100 to 200 yards" away. It was described as being a "huge creature," with its "back" visible. The husband said he "detected [seeing] a backbone...running the length of the beast." The man described "the length of the back to be of the order of 20 ft." and about "2 ft. out of the water." The animal was moving at about "10 miles per hour." Later, a "second hump" appeared. This sighting of Lizzie lasted about 1 ½ to 2 minutes (Sargent 1975).

At the lay by pull off, Pat and I ate a bag lunch as we took in the picturesque scenery. With our binoculars, we scanned the water's surface hoping to spot one of its shy denizens.

It was then on to Glenfinnan, a tranquil community on the shores of Loch Shiel, a waterway near the west coast of Scotland. When we arrived in the afternoon, it was slightly hazy and was also quite warm for that time of the year. We checked into our lodging for the next several days, the Glenfinnan House Hotel. We selected this inn

18.1 A hazy day in 1984 at the village of Glenfinnan on Loch Shiel in western Scotland. (Credit: Pat Meaney)

because it was the business and home of Tearlach (aka Charlie) MacFarlane and his family, as well as it being a stately hotel located on the eastern edge of Loch Shiel. In my journal, I would later describe our host, the cordial Charlie MacFarlane: "He is fit and with a mustache…a fine fellow" (Zarzynski 1984 journal).

The Glenfinnan House Hotel was constructed in the 1750s. The structure previously served as a hotel, a farmer's dwelling, and even a residence for a local priest. In 1971, the MacFarlane family acquired the building, which by then had fallen into some disrepair. With lots of sweat equity, the MacFarlanes gradually restored the historic structure, transforming it into a magnificent inn. The lodge had a lengthy lawn leading down to its idyllic setting—Loch Shiel.

The Glenfinnan Monument, built in 1815, was only a few-hundred yards away. It commemorates the Scots who fought and perished during the 1745 to 1746 campaign supporting Prince Charles Edward Stuart. The impressive stone pillar is about 60 feet in height including the statue of a kilted Highlander, The memorial is surrounded by an eight-sided stonewall. Several-hundred yards beyond the Glenfinnan Monument is the Glenfinnan Viaduct of the West Highland Line railway, an eye-catching bridge of 21 semi-circular stone spans, each about 50-feet tall (Visit Scotland:2019b).

Within minutes of arriving at Glenfinnan and with plenty of time before dinner, Pat and I seized the opportunity to walk around. We wanted to behold the cultural landscape of stunning Loch Shiel. There was a forestry road overlooking the waterway, but that was a fair distance away by foot. So, we just wandered around a couple of roads, taking in the exquisite scenery. Remarkably, following a 28.5-mile run the previous day, I didn't feel too tired for an extended stroll. Following our two-hour walk near the east end of the loch, we returned to the historic inn. We went out for dinner, knowing we would be taking most of our meals at the Glenfinnan House Hotel.

At the time of our 1984 visit, the tiny hamlet of Glenfinnan only had about 60 to 70 residents. Glenfinnan is nonetheless important as it is on the West Highland Line railway, running from Glasgow to Fort William, on past Glenfinnan, and then to Mallaig near Loch Morar; another branch of this railway line goes from Glasgow to Oban. Because of the remoteness of Loch Shiel, the possibility of uncovering documentation of more sightings of Seileag was rather remote back in the mid-

1980s. There was no-paved two-lane public road along the length of Loch Shiel like route A82 at Loch Ness. Rather, at Loch Shiel there was just a 12-mile road, unpaved and extremely bumpy, that ran along part of the waterway. That forestry road was shut off to public transportation (Zarzynski 1984b:50-52). I thought if Loch Shiel was a sanctuary for water monsters, few sightings would be recorded due to the lack of people gazing out over its waters.

During our August 23–29, 1984 visit to Loch Shiel, Pat and I nonetheless undertook research on Seileag sightings. We talked to numerous people, trying to glean previously unrecorded information. Likewise, Pat and I did plenty of shore watching for the mystery animals "armed" with our binoculars and cameras.

The sightings of Seileag were originally compiled over 1933 to 1934 by Dom Cyril Dieckhoff, a cleric at the Benedictine Abbey at Fort Augustus. Also, Constance Whyte, who wrote an early book on Nessie entitled *More Than a Legend: The Story of the Loch Ness Monster* (Hamish Hamilton, 1957), as well as Charlie MacFarlane, the proprietor of the Glenfinnan Hotel, added to Dom Cyril Dieckhoff's compilation of Seileag sightings. The year 1933, marked the beginning of the flap of sightings of unidentified animals at Loch Ness which must have inspired Dom Cyril Dieckhoff to dig up eyewitness accounts of Nessie-like sightings from other Scottish locales.

The following is a list of the sightings of Seileag up to 1984. This collection is from an article I wrote published in 1984 in *Cryptozoology*, the scientific journal of the International Society of Cryptozoology (Zarzynski 1984b:50-54). It gives credence to why Pat and I journeyed to that loch. We wanted to find out if there was any validity that Loch Shiel was the habitat of unidentified creatures similar to those reported at Loch Ness and Lake Champlain:

1. Circa 1874, a woman who lived at Gaskan [mid-loch, north shore] observed an animal with three humps; it moved very fast (Whyte 1957).

2. In 1911, an animal was spotted by two men across from Gaskan. Using a telescope, they saw three humps of a strange animal, each hump was separated by water. One of the witnesses was the former-head keeper of the Inverailort Estate (Whyte 1957).

3. In 1905, Ewan MacIntosh, two young boys, and an elderly man named Ian Crookback were out on the steamer *Clan Ranald* across from Gaskan. Using a telescope they spotted three humps (Whyte 1957, Costello 1974).

4. In 1925, the animal was seen at Rhu Ghainmheach. Three humps were observed through a telescope, the middle hump being the largest. The creature was described as being "longer than the little mail steamer *Clan Ranald*" (Whyte 1957).

5. In 1926, Ronald MacLeod saw a strange animal coming out of the water at Sandy Point, between 3:00 p.m. and 4:00 p.m. MacLeod studied it through his telescope; he informed his sister, Ann Mor MacDonald, that it was bigger than the steamer *Clan Ranald*, with a long and thick neck, a broad head with a wide mouth, and seven "sails" on its back (Whyte 1957, Costello 1974).

6. In December 1933, Father Dieckhoff was informed by an elderly man who lived by the Shiel bridges of a sighting of Seileag near Dalilea [about three miles from the southwest end of Loch Shiel]—date not recorded (Whyte 1957).

7. Sandie MacKellaig tried to shoot an animal (described as being a Seileag) from his boat while giving passage to two women. The women admonished him saying, "Do not interfere with it, it has not done you any harm"—no date recorded (Whyte 1957, Costello 1974).

8. In the 1950s, an unidentified animal was sighted about six miles from Glenfinnan (MacFarlane, personal communication).

9. In July 1979, there was a sighting by a pair of local youths. They were at the pier at Glenfinnan (MacFarlane, personal communication) (Zarzynski 1984:52-53).

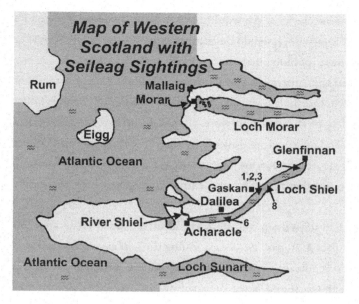

18.2 *Map of Western Scotland with Seileag Sightings; sighting numbers 4, 5, and 7 were not plotted due to the lack of specific locations provided by eyewitness(es). (Credit: Joseph W. Zarzynski)*

Friday—August 24, 1984:

Pat and I woke up early and took breakfast at the Glenfinnan House Hotel. As we waited for the dining room to open, Charlie MacFarlane, the hotel proprietor, played his bagpipes calling guests to the morning meal. What a great way to start one's day as any mental cobwebs were completely vanquished by the sound of the "tartan octopus."

After breakfast we drove over to the Forestry Commission road, parked, and hiked along the dirt roadway, much of it close to the south side of the waterway. About a mile-and-a-half into our walking tour and in sight of the loch, we spotted Charlie MacFarlane sailing his small wooden craft on the loch. Our hotel proprietor was transporting a guest across the water to near where we were hiking. The hotel patron was an Australian woman who wanted to visit the house where her great grandmother had lived many years ago. The older lady was visibly excited about this, an ancestral pilgrimage that she had wanted to take for many years.

Pat and I carried our binoculars and cameras as we walked. Pat had her Nikkormat camera with a 28/85 mm lens and I toted my Minolta XGM camera with an 80/205 mm zoom lens. We left our longer-range camera lenses at home due to their bulky size and weight. We would have to be really close if we had a sighting of a Seileag. Otherwise, it would be another one of those maybe-monster photographs of a nondescript blob on a watery surface.

We later observed the Australian woman near the house ruins of her great grandmother. The structure had been abandoned and had pretty much fallen down. All that was left of the building were chunks of structural debris. She approached us and asked me to climb down from the forestry byway and collect a stone from the dilapidated house foundation. I gladly agreed as the woman's age prevented her from doing that. She was extremely pleased with the newly acquired keepsake.

My fiancée and I had started our trek along the forestry road at 10:15 a.m. and we ended our long walk at 4:30 p.m. Besides scanning the surface waters in search of Seileag, Pat and I were likewise interested in what people we might see in this rural area, both upon the loch and along the roadway. In over six hours, we had observed only a few boats, mostly out fishing on Loch Shiel, and just a mail van, two lorries, and a single car driving along the dirt and gravel road. When we completed our hike along the loch's shoreline, I thought I sure was putting in a lot of miles on the roads of Scotland. First, some jogging around Drumnadrochit prior to August 22, then my ultramarathon, and later a lengthy walk undertaken two days after the distance run.

That night we ate a glorious dinner in the Glenfinnan House Hotel. I previously had talked to Charlie and his wife about my diet. A vegetarian, I preferred garden salads as my entrée, and they amazingly met that request. Pat and I then retired for the evening. This was a trip where I had no problem sleeping as our activities each day tired us out.

Saturday—August 25, 1984:

I awoke early that Saturday morning before breakfast, but Pat slept in a bit. I headed down to the water's edge of Loch Shiel with my camera and binoculars. The loch's surface was dead calm, ideal for watching for lake monsters. However, the region's midges, tiny flying insects, were all around me. After about 25 minutes of tolerating these troublesome bugs, I sprinted back to the hotel for breakfast.

After our morning meal, we had an hour-and-a-half long drive to Drumnadrochit to attend the Glenurquhart Highland Games that were being contested at a village field. On our trip, we stopped at Loch Lochy, Loch Oich, and Loch Ness, each time conducting some shore watching, surveying the waters for the ever-elusive beasties. Sometimes, there was something quite peaceful when searching for lake monsters. Call it a type of Zen. It sure was relaxing to one's psyche.

The first part of the Scottish games at Drumnadrochit were local events and then came the competitive rounds. At the event we bumped into Ronnie Bremner, since his Loch Ness Centre & Exhibition sponsored the hammer-throw competition. After taking in the enjoyable festivity, we drove back to Glenfinnan for dinner at the hotel.

Sunday—August 26, 1984:

I was up early in the morning to take a jog before breakfast. My legs were still tight from the ultramarathon four days earlier, so I only ran 2.1 miles. Nevertheless, I was nearly recovered from the physical exertion of my ultramarathon.

Following breakfast, Pat and I went to the nearby Church of Saint Mary and Saint Finnan in Glenfinnan for Sunday service. Later, we toured the Glenfinnan Memorial and then we ambled over to the Glenfinnan Pier to conduct shore watching for Seileag. Following our fieldwork, we drove to Loch Morar, 35 miles to the west, to visit Mary and Jim Penny at their residence.

Our stopover with Mary and Jim lasted for 1 hour and 40 minutes. Although my first trip to Loch Morar was in 1979, Pat and I had initially met the Pennys in 1982, during our cryptozoological-inspired reconnaissance of several Scottish lochs. Jim Penny was Loch Morar's water bailiff who was a veteran of the Special Air Service (SAS) of the British military. Jim had been in the Black Watch regiment. The Black Watch's lineage can be traced to the 42nd Regiment of Foot, the British army unit that was nearly wiped out at Fort Carillon (later known as Fort Ticonderoga) in Ticonderoga, New York in July 1758, during the French & Indian War (1755–1763). In Britain this conflict is known as the Seven Years' War (1756–1763). Jim Penny was anxious to hear what was new at the Fort Ticonderoga museum and its interpretation of the history of that site. Since I lived about an hour-and-a-half drive from the fort, located on the south shores of Lake Champlain, I kept abreast of Fort Ticonderoga's cultural programs. I was happy to share with the Pennys, news of the tourist and educational programs at Fort Ticonderoga.

Jim, who looked like a professional rugby player, said the warm weather during the summer of 1984 had driven more people to the cooler temperatures of the Highlands. Thus, his locality had more visitors than normal.

Reports of people seeing Morag dated to at least 1887 (Zarzynski 1984c:2). In modern times, these aquatic creatures had sparked a spectacular media splash in August 1969. William Simpson and Duncan McDonnell, two Highlanders, claimed their fishing boat had actually been rammed by a Morag beastie (Campbell and Solomon 1972:13).

A few months after that encounter, some Nessie searchers were attending a Christmas party. After much discussion, they decided to send some of their people to investigate known Morag sightings and to conduct preliminary field-work at the more-isolated Loch Morar. That was the genesis for a new-research unit that was established in 1970, and it became known as the Loch Morar Survey (LMS). The squad included biologists from London University, led by Dr. David Solomon, as well as enthusiastic volunteers. The LMS was organized into three sections: biological, operational, and historical. Several Loch Ness monster-hunting veterans likewise became involved in the LMS (Campbell and Solomon 1972:13-19). One of the reasons why so little information about Morag sightings had not filtered out from the Scottish Highlands was due to local super-stition about the "each-uisge" (water horse). Some Highlanders thought the water horses were "in league with the devil" (Campbell and Solomon 1972:23). According to this deep-seated folklore, it was considered bad luck to even talk about this phenomenon.

After returning to Loch Shiel, I found Charlie MacFarlane on the hotel grounds. He was not too busy so Charlie gave me a primer about the local lore and history of Glenfinnan. I came away astonished from the day's visit to Loch Morar and my chat with Charlie about the entrenched influence of local folklore about water horses upon the region's communities.

Monday—August 27, 1984:

After breakfast, Pat and I drove east to Corpach on Loch Eil, not far from Fort William. There we cashed some Traveler's checks and stopped at a post office to purchase stamps for postcards. At a village market we purchased some fruit, snacks, and drinks for our daily travels for the next few days.

We returned to our hotel at 11:00 a.m. and were greeted with an unexpected surprise. Charlie MacFarlane offered to give us a boat ride in his wooden sailing craft that was moored on the loch. It was a scaled-down version of a Norse-faering vessel, an open watercraft with clinker- or lapstrake-hull planks. We had observed the boat on the waterway three days earlier.

The faering craft got its name because it traditionally had four oars, as the word faering means "four oaring." This Viking-type boat's side boards, called strakes, overlapped. This hull construction is known as clinker or lapstrake planking. The strakes were fitted together with metal rivets rather than nails. Small vessels of this size are typically carvel planked, that is, the hull's side planks are edge-to-edge. Charlie's boat was well-constructed and it could be rowed or sailed.

18.3 Replica of a Norse faering-like vessel at Loch Shiel, one of the waterways in Scotland with a history of monster sightings. (Credit: Pat Meaney)

There was a slight breeze upon the loch during our cruise. Because of the skill of our pilot, we sailed back to shore, rather than employ any oars. With expertise, Charlie landed his boat right in front of the hotel. We had no sighting of Seileag during the excursion, but the sailboat voyage with our expert mariner was a thrill. It was times like this that Pat and I realized how fortunate we were to be doing things like chasing reputed lake monsters, running ultramarathons along fabled Loch Ness, visiting an earthen mound where a slain dragon was said to be buried, and sailing in a replica Viking craft on fabled Loch Shiel in the Highlands.

Afterwards, I was able to get in an easy three-mile jog. I had finally gotten back my "running legs," though one knee was sore from the ultramarathon five days earlier.

After dinner, Pat and I went to the hotel bar for a ceilidh, a traditional Scottish gathering with music, dance, drinking, and friendship. Charlie was on the bagpipes while accompanied by a neighbor on the fiddle. It was a rousing celebration of the cultural mores of the region.

Tuesday—August 28, 1984:

It was raining in the morning, so we drove around for the day doing some sightseeing. We first traveled to Acharacle at the southwest end of Loch Shiel. We walked to a local tea room for coffee before heading east to Dalilea, a tiny community on the north shore of Loch Shiel. By then the rain had stopped, but it was damp, overcast, and somewhat chilly. Nonetheless, we hiked for two hours along Loch Shiel, hoping to catch a glimpse of one of the shy Seileag creatures. Following our cryptozoological fieldwork, we returned to Glenfinnan. I went out for a 2.1-mile jog. After dinner we again talked to our amiable host, Charlie MacFarlane, and gained further insight into the area's history and lore before retiring for the evening. Tomorrow, we would be leaving Loch Shiel. We would miss this alluring locale near the west coast of Scotland and its friendly residents.

Wednesday—August 29, 1984:

Pat and I departed the Glenfinnan House Hotel after a breakfast prepared early in the morning by Charlie and his delightful wife. The couple was so gracious to us

18.4 Lionel Leslie, a retired British Army officer, adventurer, and hunter of lake monsters, at his house on the Isle of Mull, Scotland in 1984. (Credit: Pat Meaney)

during our stay investigating Seileag. It was then on to Lochaline, an automobile trip of a little over one-and-a-half hours, to catch the car ferry to the Isle of Mull.

We journeyed to Mull to visit Lionel Leslie, a veteran of several cryptozoological projects in both Scotland and Ireland. Lionel and his wife Barbara lived at Grasspoint, a remote spot on the east side of the island. The couple ran a cozy tea room where they also sold some of their art. Lionel had been an officer in the British Army. Captain Lionel Leslie was somewhat unique in the field of cryptozoology, since he had studied some of the loughs (Irish spelling for loch) in Ireland that had a reputation of having Nessie-like animals.

Nearly three years after our visit to the Leslies, we were notified that Lionel had passed away in 1987. His obituary, published in *The Times* (London) on January 21, 1987, referred to him as a sculptor, soldier, and explorer. Lionel Leslie served in the British Army in India and Burma and then left military service to explore Africa in the late 1920s. Accompanied by a Masai guide, he once attempted to walk across Africa. When I read that, it made my 28.5-mile run of Loch Ness seem quite paltry in comparison.

After departing Africa, Lionel Leslie lived in Labrador, Canada. There, he worked on a vessel engaged in rum smuggling along the Canadian Atlantic coast.

His first book was published in 1931, and it was appropriately entitled: *Wilderness Trails in Three Continents—An Account of Travel, Big Game Hunting and Exploration in India, Burma, China, East Africa and Labrador.* The non-fiction book included a forward by Winston Churchill, his cousin and godfather.

Leslie then studied sculpture in Paris for four years. Returning to the United Kingdom, he shared an art studio in London with a Russian sculptor. Several of Lionel's sculptures were exhibited in galleries and museums in Britain. During our 1984 visit with the Leslies, I was particularly impressed with Lionel's sculptures that he had for sale at his tea room at Grasspoint.

At the outbreak of the Second World War in 1939, the military veteran returned to service in the British Army. During the war in 1942, he married Barbara Enever and the couple eventually moved to the Isle of Mull. In 1961, Lionel Leslie published his autobiography entitled *One Man's World: A Story of Strange People and Strange Places.* In the 1960s, he spent time at Loch Ness, where he experimented with nighttime surveillance of the waterway pursuing Nessie animals that he believed were nocturnal. Lionel Leslie likewise traveled to Connemara, Ireland where the monster hunter conducted fieldwork seeking lough monsters in Irish waters.

One day while Lionel was walking the beach on the jagged shores of the Isle of Mull in Scotland, he discovered the carcass of a stranded killer whale. He retrieved its skull, cleaned it, and exhibited the marine specimen in the family tea room (*The Times* 1987).

After an exhilarating visit with the Leslies, we departed the island and drove to Oban and then south to Loch Fyne, where we stayed the evening. While motoring toward Oban, Pat and I discussed our visit with the Leslies. We came away greatly impressed with Lionel's modesty and the incredible friendliness that both Barbara and he bestowed upon two strangers from distant upstate New York.

Thursday—August 30, 1984:

After breakfast at our hotel near Loch Fyne, Pat and I drove to Glasgow where we shopped for gifts for our families and friends. We stayed the evening in the Seamill, Scotland area, north of Prestwick, in the same hotel where we spent our first evening during this long and eventful expedition to Scotland over the summer of 1984.

Friday—August 31, 1984:

After our morning meal we motored to Prestwick Airport, where we had landed to start our two-week trip through the Scottish Highlands. After a delay of a few hours, what else was new on this trip, our aircraft finally departed the United Kingdom, heading to the United States. We certainly had a most-memorable trip since we departed our three-room apartment in the early morning of August 16, 1984.

1984 Summary:

The year 1984 was a busy one for our cryptozoological fieldwork in Scotland and also at Lake Champlain. Due to the lengthy duration of our expedition to Scotland, my research entity, known as the Lake Champlain Phenomena Investigation, conducted a scaled-down season of summertime monster hunting for Champ. The majority of our efforts at Lake Champlain were based from a lakeside cabin near Vergennes, Vermont, a place we had rented each summer since 1981 to search for Champ.

In 1984, Pat and I conducted 17 days of Champ-related fieldwork at Lake Champlain which included shore- and waterborne-based camera work, the latter from my 13-foot-long inflatable boat. Furthermore, we undertook scuba reconnaissance looking for a Champ carcass. Our enthusiastic team that year included my

18.5 Divers Zarzynski and Meaney with their sonar-equipped tripod at Lake Champlain, searching for the maybe-monsters nicknamed Champ. (Credit: Joseph W. Zarzynski Collection)

fiancée Pat Meaney, Ted Straiton, Rod Canham (owner of a scuba shop in Johnson City, New York), and myself (Zarzynski 1984a:80-83). Moreover, on August 9, 1984, Pat and I assisted cryptozoologist Richard D. Smith (Wind & Whalebone Media Productions) during his four-day underwater videography trials at Lake Champlain (Zarzynski 1984a:80). In the 1980s, the Princeton, New Jersey resident became intrigued with cryptozoology. He employed an underwater video camera that was towed from a motorboat, attempting to collect video documentation to solve the Champ puzzle (Smith 1984:89-93). Smith's fieldwork was assisted by several people including Connecticut resident Gary Mangiacopra. Gary was a dedicated archivist whose voluminous files on sea monsters were the most comprehensive in the country.

In 1984, Pat and I had spent a month conducting fieldwork in Scotland and Lake Champlain. At the time, I did not realize that in the following year, 1985, I would begin to redirect my exploration efforts into another discipline. The catalyst for that would be a dramatic salvage operation that occurred in September 1985—the recovery of a rare World War II Wellington bomber from the peat-stained depths of Loch Ness.

Chapter 19

After Cryptozoology—
My Underwater Archaeology Career

. . . .

Along with being a school teacher in the Saratoga Springs City School District for 31 years, I had two other lengthy research careers. The first was being a cryptozoologist at Loch Ness, Lake Champlain, and elsewhere. My other major interest, actually longer in duration than my cryptozoological pursuits, has been in the field of underwater archaeology.

My introduction into exploring sunken vessels started in 1980 at Lake Champlain. Two acquaintances, Scott Hill and Jim Kennard, both Rochester, New York-area ship-wreck explorers, were helping my fieldwork looking for Champ. Jim had a side scan sonar and he offered his remote sensing gear to scan the depths of Lake Champlain searching for a Champ carcass while he also hoped to find sunken vessels. In return, I did research for him on the shipwrecks and sunken aircraft in Lake Champlain. Jim Kennard would go on to have a storied career as a shipwreck hunter, discovering many sunken ships, especially in Lake Ontario (Kennard, Stevens, and Pawlowski 2019).

In 1875, the 258-foot-long steamboat *Champlain II* ran aground along the shoreline of Lake Champlain, two miles north of Westport, New York (Zarzynski 1986:15-18). On September 11, 1980, we located the shipwreck using Kennard's sonar. Hill and Kennard then loaned me scuba gear and I made a cursory dive on the sunken steamer. I barely had time to glimpse the shallow-water end of the wooden shipwreck that was in 10 feet of water, before I surfaced. That abrupt baptism into scuba convinced me to take a course to get certified as a scuba diver. That opened a new door of inquiry for me.

I soon began diving with Jack Sullivan, a veteran scuba instructor who resided in my neighborhood. We investigated the depths of Lake Champlain and Lake George, the latter only 20 miles from my apartment.

Because I was a teacher, I had a 10-week summer vacation. That allowed me plenty of time to pursue cryptozoology and later, underwater archaeology. Without the former, cryptozoology, I probably would not have migrated into underwater archaeology, or as it is sometimes known, maritime archaeology. My second master's degree, that in archaeology and heritage, was acquired in 2001, and I put that "sheepskin" to proper use. Today, I have pretty much retired from scuba diving after having made 2,768 open-water dives, a lot of time spent aboard boats and being underwater.

From 1974 into the year 1991, I devoted most of my spare hours undertaking cryptozoological research and fieldwork at Loch Ness, Lake Champlain, and elsewhere. For a period of about four years, from 1987 to 1991, I actually attempted to balance both of my interests—cryptozoology and underwater archaeology.

19.1 Underwater archaeologist Zarzynski collecting a measurement at the bow of the 1758 Land Tortoise *radeau shipwreck in Lake George in 1991. The author led the team that used a Klein side scan sonar to find the British vessel in 1990. The sunken vessel is known as "North America's oldest intact warship." (Credit: Dr. Russell P. Bellico)*

Eventually, by 1991, I turned my full avocational attention to underwater archaeology. Yet, I never lost interest in cryptozoology.

What was the spark for my leap into underwater archaeology? To a major extent it was Marty Klein's and Charlie Finkelstein's 1976 discovery of "R for Robert," the Loch Ness Wellington bomber, and then the plane's recovery from the Scottish waterway in 1985. I journeyed to Loch Ness in September 1985 when that World War II aircraft was lifted from its resting place of 45 years. N2980, as the aircraft was known, was a veteran of 14 combat missions. This was during the early months of the war, when the average number of combat missions for this class of warplane was only six (The Loch Ness Wellington Association Ltd. 1985:n.p.). Using primarily an atmospheric diving suit (ADS),* supplemented by scuba personnel, the historic aircraft was recovered from about 230 feet of water, an underwater salvage that totally inspired me.

19.2 An atmospheric diving suit (ADS) after making a 1985 descent to a sunken World War II plane lying in Loch Ness. (Credit: Joseph W. Zarzynski)

*An atmospheric dive suit (ADS) resembles a suit of armor with a single person in the anthropomorphic submersible. Oceaneering International, Inc. had two ADSs for their 1985 project. One was a Wasp and the other, a Jim suit. Both allowed the "diver" inside to be at the same atmosphere as a person on land (Oceaneering International, Inc. n.d.). The more-sophisticated Wasp had articulated arms, mechanical tools as "hands," and thrusters for movement. The Wasp and its diver affixed a metal lifting frame, with air bags, over the sunken-World War II bomber. When the flotation was filled with air, it raised the sunken plane to the surface.

Beginning in the mid-1980s, I began collaborating on shipwreck projects with Dr. Russell P. Bellico, a professor of economic history at Westfield State College in Massachusetts. Russ and his wife Jane resided in Massachusetts, but they were also summer residents at Lake George, a 32-mile-long waterway located 60-miles north of Albany, New York. Russ and I made dozens of scuba dives on shipwrecks in Lake Champlain, Lake George, and in New Hampshire's Lake Winnipesaukee. We soon developed a plan to create a shipwreck preserve at Lake George, literally an underwater park for scuba buffs. Since the small state of Vermont had successfully established a state-administered shipwreck park for divers in 1985 (Cohn 2003:86), Russ and I thought, why not in New York, the self-designated "Empire State."

Russ Bellico was an experienced diver and he was one of the best underwater photographers around. The Connecticut native was likewise an accomplished Lake George and Lake Champlain maritime and military historian.* Our viewpoint was that Lake George warranted an intensive study and inventory of its collection of historic shipwrecks and other notable submerged cultural resources.** Lake George, located south of Lake Champlain, had a bountiful history related to: early indigenous peoples (aka Native Americans), the French & Indian War (1755–1763), the Revolutionary War (1775–1783), the age of waterborne-steam transportation, naphtha watercraft, twentieth-century gasoline-powered vessels, and the development of nineteenth- and twentieth-century recreational and cultural tourism (Bellico 2001).

By 1987, our efforts had finally gained some momentum. With valuable assistance from two underwater explorers—Vince Capone and Jack Sullivan—we

---

*Bellico wrote several well-received books—*Sails and Steam in the Mountains: A Maritime and Military History of Lake George and Lake Champlain* (Purple Mountain Press, 1992; second edition published in 2001), *Chronicles of Lake George: Journeys in War and Peace* (Purple Mountain Press, 1995), *Chronicles of Lake Champlain: Journeys in War and Peace* (Purple Mountain Press, 1999), and *Life on a Canal Boat: The Journals of Theodore D. Bartley, 1861–1889* (Purple Mountain Press, 2004). The latter book was edited by Bellico and the Preface and Postscript were written by Arthur B. Cohn (Purple Mountain Press, 2004). Bellico's most recent book is *Empires in the Mountains: French and Indian War Campaigns and Forts in the Lake Champlain, Lake George, and Hudson River Corridor* (Purple Mountain Press, 2010).

**Submerged cultural resources can include aircraft, artifacts, bridges, buildings, vessels, and wharves that are in an underwater environment.

initiated a strategy to establish a shipwreck park for scuba divers at historic Lake George. We organized a three-day underwater-archaeology training session at Lake George in 1987, that resulted in the formation of a small, but enthusiastic organization called the Atlantic Alliance Lake George Bateaux Research Team. That entity was later renamed the Lake George Bateaux Research Team. The all-volunteer group eventually evolved into the not-for-profit organization—Bateaux Below, Inc.* (Zarzynski 2019:73-91).

Bateaux Below consisted of six principals—Dr. Russell P. Bellico, Bob Benway, Vince Capone, Terry Crandall, John Farrell, and myself. The organization was active for nearly a quarter-of-a-century from 1987 until 2011. We had energetic support, too, from tireless volunteers such as maritime archaeologist Dr. D.K. (Kathy) Abbass and scuba divers Bill Appling, Dr. Sam Bowser, Steve Cernak, Paul Cornell, Bob Leombruno, Maria Macri, Kendrick McMahan, Scott Padeni, Mark L. Peckham, Peter Pepe, Steve Resler, and David Van Aken, to name a few. I served as Bateaux Below's executive director.

A few of our successes include:

1. From 1987 to 1991, our archaeological team mapped seven 1758 British shipwrecks in Lake George, known as the Wiawaka Bateaux

2. During a 1987 to 2011 submerged cultural resources inventory, we employed Klein side scan sonar, supplemented by scuba diver reconnaissance, to catalogue over 200 shipwrecks in Lake George

---

*For more information about the underwater archaeological and public outreach work of the Lake George Bateaux Research Team and Bateaux Below, Inc., see my books: *Ghost Fleet Awakened: Lake George's Sunken Bateaux of 1758* (SUNY Press, 2019); *Lake George Shipwrecks and Sunken History*, co-written with Bob Benway (The History Press, 2011); and *Documentary Filmmaking for Archaeologists*, co-authored by Peter Pepe (Left Coast Press, 2012). Also, see the DVD documentaries: *The Lost Radeau: North America's Oldest Intact Warship* by Pepe Productions, Bateaux Below, Inc., Whitesel Graphics, and Black Laser Learning (2005, 57 minutes) and *Wooden Bones: The Sunken Fleet of 1758* by Pepe Productions and Bateaux Below, Inc. (2010, 50 minutes).

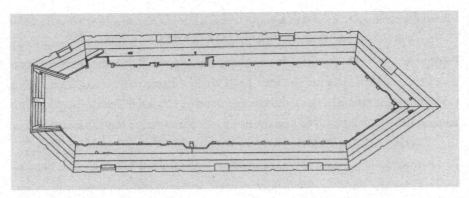

19.3 A drawing, in plan view, of Lake George, New York's 1758 Land Tortoise *radeau shipwreck. The seven-sided wooden warship, a type of floating battery, is 52 feet long × 18 feet wide and was pierced for seven cannons. Following a career in cryptozoology, the author was project manager on the archaeological mapping of this deepwater British warship from the French & Indian War (1755–1763). (Credit: Kerry Dixon and Bateaux Below, Inc.)*

3. In 1990, using a Klein side scan sonar, we discovered the 1758 *Land Tortoise* radeau shipwreck, known as "North America's oldest intact warship." From 1990 to 1994, we mapped the deepwater shipwreck

4. From 1993 to 1994, we created the first seamless photomosaic of a shipwreck, Lake George's 1758 *Land Tortoise* radeau, with the underwater photography by Bob Benway and the computer assemblage of the photomosaic by Kendrick McMahan

5. In 1993, I was co-operator of a Klein side scan sonar during a survey for the Rhode Island Marine Archaeology Project (RIMAP), off Newport, Rhode Island, that searched for sunken Revolutionary War naval transports and other shipwrecks

6. Bateaux Below divers assisted RIMAP on numerous archaeology projects in Rhode Island's Narragansett Bay (1993 to 2000, 2008 to 2009, and 2012 to 2015)

7. Bateaux Below divers worked with other underwater archaeologists (1990 and 1996) to map a shipwreck off the Florida Keys, hypothesized to be the 1822 wreck of the US Schooner *Alligator,* whose crew enforced the ban on the trans-Atlantic slave trade and later battled Caribbean pirates

8. Bateaux Below deployed a Klein side side sonar that discovered and later our divers mapped (1997 to 1999) the 1893-built *Cadet* ex *Olive,* a steam launch sunk in Lake George

9. From 1993 to 2000, Bateaux Below members mapped Lake George's 1910-built Delaware & Hudson marine railway that launched boats from boxcars

10. From 1998 to 2000, Bateaux Below archaeological divers worked with RIMAP to map the HMS *Cerberus,* a British Revolutionary War frigate, sunk in Narragansett Bay, Rhode Island in 1778

11. In 2000, I was one of the Klein side scan sonar operators for RIMAP's search for Captain James Cook's ship of discovery, the HMB *Endeavour,* sunk in Newport, Rhode Island's outer harbor in 1778

12. From 2002 to 2004, Bateaux Below members mapped a submerged 1758-built French & Indian War wharf in Lake George

13. From 2007 to 2008, Bateaux Below personnel advised Saratoga Springs, New York middle school students and their technology-class teachers in the construction of a 30-foot-long replica of a 1758 bateau "wreck," that was sunk in Lake George's shallows for pedestrian viewing

14. Bateaux Below was the principal creator and monitor of New York State's "Submerged Heritage Preserves," a shipwreck park in Lake George for scuba divers (1993 to 2011).

By the year 2011, core members and volunteers of Bateaux Below had crossed into their senior-citizen years. Also, one team volunteer passed away, two others developed health problems, another resigned, and one moved into aquatic invasive species management. The handwriting "was on the wall." Bateaux Below ceased its underwater archaeology projects. It became time to finish writing articles, books, and reports that we never had the many hours to complete and get published.

I had the privilege to co-found the Lake George Bateaux Research Team and also Bateaux Below. I served as director of both organizations. My passion for underwater archaeology started after gaining invaluable experience while searching for lake monsters. Life is sometimes unconventional. I went from lake monster hunter to underwater archaeologist. Remarkably, it was a career arc that worked for me.

These days, I devote my professional efforts as a volunteer for not-for-profit organizations. My wife and I assist cultural institutions, too, working on terrestrial archaeology-fieldwork projects and doing post-excavation laboratory cataloguing. Moreover, I write a regular column on history and archaeology topics for the *Lake George Mirror* newspaper, a non-paid position. I also research and write books. All that would not have happened had I not in September 1974, stumbled upon Tim Dinsdale's book—*Monster Hunt*. The hand of fate occasionally is strange, sometimes as out of the ordinary as that which you seek.

Chapter 20

The Loch Ness Marathon

In 2002, eighteen years after I ran my one-person 28.5-mile ultrarun, a Loch Ness marathon (26.2 miles) had its inaugural race. Drumnadrochit's Ronnie Bremner had a thorough understanding of cultural and recreational tourism at Loch Ness, though this wonderful marathon was not his handiwork. Rather, the sponsor is a food-processing company. The race is known as Baxters Loch Ness Marathon and it is one of the most popular marathons in Great Britain. It is contested along a section of Loch Ness, that covers about two-thirds the length of the famous waters.

20.1 Runners competing in the 2019 Baxters Loch Ness Marathon race along part of the east side of Loch Ness. (Credit: Baxters Loch Ness Marathon and Reuben Tabner)

The inaugural Baxters Loch Ness Marathon was on September 27, 2002 and it attracted over 800 competitors (Clark 2019). However, unlike my one-person run along route A82, Baxters Loch Ness Marathon is along the eastern side of the loch, across the waterway from route A82. Furthermore, the marathon starts at Whitebridge, a village that is located halfway between Fort Augustus and Foyers. Thus, the Loch Ness marathon does not traverse the full length of the waterway. My run along route A82 started at Fort Augustus, at the south end of Loch Ness, and I finished just past Lochend, at the waterway's other end. Baxters Loch Ness Marathon begins on route B862, about nine miles from Fort Augustus, and heads northeast. The competitive event then takes B852 and finishes in Bught Park in Inverness.

This marathon and its other races of lesser distances draw several-thousand eager participants each year. The races are well liked and what's more, the runners can say they ran along fabled Loch Ness. Collectively these contests are known as Baxters Loch Ness Marathon and Festival of Running. This includes not only the 26.2-mile marathon, but other running contests, too: the River Ness 5-kilome-

20.2 Lloyd Scott finishing his underwater walk of Loch Ness for charity in 2003. (Credit: Andrew Milligan and Alamy)

ter race (3.1 miles), a 10-kilometer race (6.2 miles), the 10-kilometer Corporate Challenge with teams of four to six people, and a fun run for kids known as the Wee Nessie (Baxters Loch Ness Marathon & Festival of Running 2019).

Rather unbelievably, after the first Baxters Loch Ness Marathon, Lloyd Scott, a daring and compassionate Englishman, walked the length of Loch Ness in 2003, and he did so underwater. Scott wore a 180-pound hardhat-diving suit. A leukemia survivor, this do-gooder completed the subsurface walk in 12 days, often spending several hours a day underwater. Reportedly, Lloyd Scott's long-distance submerged odyssey never went deeper than 30 feet of water. This spectacular achievement by a former British firefighter raised money and awareness for the charity—Children with Leukemia (Kelbie 2003). Bravo to Mr. Scott.

Chapter 21

Conclusion

· ·

Looking back over 35 years ago to my 1984 ultramarathon, I am still overjoyed that I completed the lengthy Loch Ness run. I did not see a Nessie during my scamper, but I still believe there are mysteries in Loch Ness that need to be resolved.

Since the "Golden Age of Monster Hunting at Loch Ness," the late 1960s into the 1980s, the reputations of some of those that took up the challenge to investigate the zoological mystery have been somewhat sullied. That, even though there are still about 10 sightings of a Nessie each year (BBC News 2019). Over the past-three decades, a growing list of people have been critical of monster searching at Loch Ness. More recently, the results of a DNA study, completed by a team of New Zealand scientists, initially dimmed the hopes of Nessie believers.

On September 5, 2019, the BBC News published a story entitled "Loch Ness Monster may be giant eel, say scientists." Dr. Neil Gemmell, a geneticist from New Zealand's University of Otago, one of the top universities in the world, released the results of his team's DNA study at Loch Ness. The university's scientists extracted DNA from water samples taken from around Loch Ness in an attempt to catalogue the many species in the waterway. The New Zealand researchers reported they found no evidence that a prehistoric plesiosaur, one of the most popular candidates for Nessie, resided in the loch.

What the University of Otago geneticists did uncover, however, was that Loch Ness has plenty of eel DNA. The eels migrate from the Sargasso Sea in the Atlantic Ocean near the Bahamas and swim into Loch Ness, a distance of about 3,100 miles. Now that's an ultramarathon. The ray-finned fish enter the waterway via a swim up the River Ness. Dr. Gemmell's study suggested Nessie was possibly none other than an oversized eel, only a few feet in length (BBC News 2019).

That scientific analysis seemed to have sealed the fate of Nessie until a few months later. In January 2020, the *Daily Express*, a British newspaper, published the story—"Loch Ness Monster: How 'unidentified' DNA was found in Scottish Highland waters." This was a follow-up article on Dr. Neil Gemmell's DNA studies of the waters of Loch Ness. In this news account, Gemmell reported that about 20 percent of the samples of DNA extracted at Loch Ness turned out to be "unidentified," leaving the door ajar that the loch could have unidentified monstrous animals in it (Hoare 2020). Nessie hunters must have rejoiced at that news. Many probably dusted off their binoculars and cameras to renew their cryptozoological sleuthing at this world-famous waterway.

I believe I may have been the first person to run the full length of Loch Ness. Years after my 1984 ultramarathon, however, I discovered I actually have some heart issues, two that have been with me all my life. I was premature and an incubator baby, weighing only 4 pounds, 4 ounces at birth. That probably contributed to my cardiac concerns—mitral valve prolapse (a leaky heart valve) and left bundle branch block.

People with mitral valve prolapse sometimes have shortness of breath (especially during exercising), occasional dizziness, and even fainting. Left bundle branch block is a delay along the path where electrical impulses travel to the heart. This could lead to a heart attack, high blood pressure, and a thickened or weakened heart muscle. What's more, it seems that for much of my adult life I have always had a low pulse. Yet, more recently it has progressed to more than just a "runner's pulse," a low pulse due to distance running. An average healthy heart rate (resting heart rate) is between 60- and 100-beats per minute. In 2020, my low pulse, in medical jargon known as bradycardia, was recorded as low as 32-beats per minute. Down the road I probably will need a pacemaker, an electronic device surgically implanted into one's chest with wires to the heart. This medical accessory helps the heart pump correctly. Therefore, I probably was quite lucky in 1984 to have been able to complete my Loch Ness ultramarathon.

Distance runners are a rather unusual lot. Even today in 2021, and at 70 years of age, I still run six to seven days a week and participate in half-marathon events. I no longer have much speed, but I do enjoy running. As frequently uttered by jog-gers, "Once a runner, always a runner."

During my years in cryptozoology, I occasionally had people point out to me, "Hey, you're wasting your time and money with that pseudo-science stuff." I never saw it that way. Not then, nor today. Lake monster investigating was a grand adventure. It likewise was an inquiry into something special and important to me—finding the truth. Besides, during 17 years as a lake monster hunter chasing clues to the enigmas of Loch Ness and Lake Champlain, I met many enlightened, intelligent, and genuine people. Therefore, being a cryptozoologist enriched my life and that of my wife, too.

I once read, "Life is all about the journey." If so, my solo run of Loch Ness in 1984 was a metaphor of my time on earth, and all in 4 hours, 23 minutes, and 26 seconds. And it was rather ironic, too, that the finish line of my August 22, 1984 ultramarathon was none other than at a place called Lochend.

Almost three-quarters of planet Earth is covered by water—oceans, seas, lakes, ponds, rivers, streams, and canals. In my home state of New York, there are about 8,000 waterways (Zarzynski 2019:36). Yet, humanity still has mapped only about 20 percent of the world's waters. This inner space is surely filled with bewildering secrets awaiting explanation. I believe there are still new species of aquatic life yet to be discovered by bold pathfinders. Just maybe, extinct species like marine plesio-saurs or the ancient whale called basilosaurus (aka zeuglodon), may have survived into the twenty-first century.

There is an intriguing by-product of cryptozoology, one that seldom is ac-knowledged. You only need to look at Marty Klein's and Charlie Finkelstein's 1976 discovery of the Loch Ness Wellington bomber, a sensational find made during a remote sensing survey searching for water monsters. Finding that rare World War II military aircraft definitely increased "the fund of human knowledge," adding to the record of aviation history.

In 1986, the year after the recovery of the Loch Ness Wellington bomber from Loch Ness, "Doc" Harold Edgerton, the renowned-MIT professor and a member of the Academy of Applied Science, wrote to the University of Edinburgh's Dr. Ian Morrison. The pair had collaborated on some research work at Loch Ness. Edger-ton commented: "Many factors point to no 'Nessie.' Regardless, there is no harm in looking, especially with sonar since there may be things to discover" (Naone 2007).

21.1 "Doc" Harold Edgerton, a pioneer in the development of sonar, operates a Klein side scan sonar. Edgerton was one of several notable explorers that spent time using high-tech gear at Loch Ness in search of Nessie. (Credit: Martin Klein)

I wholeheartedly agree with Edgerton's belief in taking chances and optimizing opportunities for the sake of possibly making great discoveries. From my perspective, we could use more puzzles like the Loch Ness monsters to track.

My Loch Ness ultramarathon was the most exciting run I ever did. For me, the mid-1970s into the mid-1980s was a decade of escape into Scotland's "Lost World" of maybe-monsters. Even for one who once "chased" denizens of the deep in a pair of running shoes. Nessie, you're still the "hide-and-seek champion since 1933." I really can't say for sure if the Loch Ness cryptids exist. But possibly, one day we may find out all of Loch Ness' secrets.

21.2 A full-scale fiberglass model of a Nessie creature floating in a Drumnadrochit, Scotland pond. (Credit: Pat Meaney)

Bibliography

Amdur, Neil. "2 World Records Set as 13,360 Finish Marathon." *New York Times,* October 26, 1981.

American Battlefield Trust. "Sybil Ludington." Washington, DC: American Battlefield Trust. Accessed January 27, 2020. www.battlefields.org/learn/ biographies/sybil-ludington.

Associated Press. "Shadows in Loch Ness." *The Herald Statesman* (Yonkers, NY), November 7, 1976.

———. "Dolphin for 'Nessie' hunt dies." *Binghamton Press* (Binghamton, NY), June 27, 1979.

Bauer, Henry H. *The Enigma of Loch Ness—Making Sense of a Mystery.* Urbana, IL: University of Illinois, 1988.

Baxters Loch Ness Marathon & Festival of Running. Accessed January 22, 2019. www.facebook.com/pg/lochnessmarathon/about/?ref=page_internal.

BBC News (London). "Loch Ness Monster may be giant eel, say scientists." September 5, 2019. Accessed January 22, 2020. www.bbc.com/news/uk-scotland-highlands-islands-49495145.

Bellico, Russell P. *Chronicles of Lake Champlain: Journeys in War and Peace.* Fleischmanns, Purple Mountain Press, 1999.

———. *Chronicles of Lake George: Journeys in War and Peace.* Fleischmanns, NY: Purple Mountain Press, 1995.

———. *Empires in the Mountains: French and Indian War Campaigns and Forts in the Lake Champlain, Lake George, and Hudson River Corridor.* Fleischmanns, Purple Mountain Press, 2010.

———. *Sails and Steam in the Mountains: A Maritime and Military History of Lake George and Lake Champlain.* Fleischmanns, NY: Purple Mountain Press, 1992, (1st ed.).

———. *Sails and Steam in the Mountains: A Maritime and Military History of Lake George and Lake Champlain.* Fleischmanns, NY: Purple Mountain Press, 2001, (revised ed.).

Bellico, Russell P. (ed.). *Life on a Canal Boat: The Journals of Theodore D. Bartley 1861–1889.* Fleischmanns, Purple Mountain Press, 2004.

The Benleva. "The Benleva." Accessed September 28, 2019. http://www.benleva.co.uk.

Binghamton Evening Press (Binghamton, NY). "Finish St. Paul's High School Courses." July 1, 1937.

Blundell, Rev. Odo. "Notice of the Examination, by means of a Diving-dress of the Artificial Island, or Crannog, of Eileen Muireach, in the South End of Loch Ness." *Proceedings of the Society of Antiquities of Scotland One Hundred and Twenty-Ninth Session 1908–1909.* Vol. 43. pp. 159-163. Edinburgh: Neill and Company Ltd., 1909.

Breyer, Lucy A. "National Register of Historic Places—Saratoga Spa State Park." Albany, NY: NYS Office of Parks, Recreation and Historic Preservation, September 12, 1985.

Burfoot, Amby. "Why Did American Men Run Faster In New York Decades Ago?" *Runner's World,* November 4, 2016. Accessed April 29, 2020. www.runnersworld.com/news/a20830743/why-did-american-men-run-faster-in-new-york-decades-ago/.

Burton, Maurice. *The Elusive Monster—An Analysis of the Evidence from Loch Ness.* London: Rupert Hart-Davis, 1961.

The Caledonian Mercury (Edinburgh). "Caledonian Canal." October 29, 1821.

Campbell, Elizabeth Montgomery and David Solomon Ph.D. *The Search for Morag.* London: Tom Stacey Ltd., 1972.

Caulfield, Philip. "Elvis impersonator in Vegas runs marathon, gets married and saves woman's life all in one day." *New York Daily News,* December 8, 2010.

Clark Athletics. "2006 Clark University Athletics Hall of Fame Inductees Announced." Worcester, MA: Clark University. 2006. Accessed April 15, 2020. www.clarkathletics.com/HOF/Releases/2006.

Clark, Will. "Loch Ness Festival of Running will feature biggest ever number of entries." *Inverness Courier* (Inverness, Scotland), October 3, 2019. Accessed January28, 2020. www.inverness-courier.co.uk/sport/marathon-is-set-to-break-record-entry-183853/.

Cohn, Arthur B. "Lake Champlain's Underwater Historic Preserve System: Reasonable Access to Appropriate Sites." *Submerged Cultural Resource Management: Preserving and Interpreting Our Sunken Maritime Heritage.* pp. 85–94. James D. Spirek and Della A. Scott-Ireton, eds. New York: Kluwer Academic/Plenum Publishers, 2003.

Coleman, Loren. "International Cryptozoology Society Founded, 2016." *CryptoZooNews*, January 25. 2016. Accessed March 24, 2020. www.cryptozoonews.com/ics-2016-pt1/.

———. "Loch Ness Monster & Mokele-Mbembe Researcher, Cryptozoologist Roy P. Mackal Has Died." *CryptoZooNews*, December 15, 2013. Accessed January 5, 2020. www.cryptozoonews.com/mackal-obit/.

———. "Nessie Seeker Frank Searle Dies." *The Cryptozoologist*, 2005. Accessed January 23, 2020. www.lorencoleman.com/searle.html.

———. "Ronald 'Ronnie' Bremner (1941-2001) Owner of the Loch Ness Center [sic Centre] and Exhibition." 2003. Accessed January 28, 2019. www.lorencoleman.com/ronald_bremner_obituary.html.

Coleman, Loren and Patrick Huyghe. *The Field Guide to Lake Monsters, Sea Serpents, and Other Mystery Denizens of the Deep.* New York: Tarcher/Penguin, 2003.

Costello, Peter. *In Search of Lake Monsters.* New York: Coward, McCann & Geoghegan, 1974.

Crockett, Davy. "Yiannis Kouros—Greek Greatness." UltraRunning History—Podcast and Stories—Davy Crockett, 2019. Accessed May 9, 2020. www.ultrarunninghistory.com/yiannis-kouros/.

Daly, Mark. "Abuse at the Abbey: How paedophile [pedophile] monks were finally exposed." *The Independent* (London), August 13, 2013. Accessed January 23, 2020. www.independent.co.uk/news/uk/crime/abuse-at-the-abbey-how-paedophile-monks-were-finally-exposed-8773459.html.

Denby, David. "D Day on Film." New York: *New Yorker,* June 6, 2014. Accessed April 4, 2020. www.newyorker.com/culture/culture-desk/d-day-on-film.

Dick's Sports. "1985 Overlook Overload." Kingston, New York: Dick's Sports, 1985.

Dinsdale, Angus. *The Man Who Filmed Nessie: Tim Dinsdale and the Enigma of Loch Ness.* Surrey, BC, Canada: Hancock House Publishers Ltd., 2013.

Dinsdale, Tim. Letter to author. December 4, 1984.

---. *The Leviathans.* London: Routledge & Kegan Paul Ltd., 1966.

---. *Loch Ness Monster.* London: Routledge & Kegan Paul Ltd., 1961.

---. *Monster Hunt.* Washington, DC: Acropolis Books Ltd., 1972.

---. *Project Water Horse: The True Story of the Monster Quest at Loch Ness.* London: Routledge & Kegan Paul Ltd., 1975.

---. *The Story of the Loch Ness Monster.* London: Target Books, 1973.

Doyle, A. Conan. *The Lost World.* London: Hodder & Stoughton, 1912.

Ellis, William S. "Loch Ness—The Lake and the Legend." *National Geographic Magazine.* pp. 758–779. Washington, DC: National Geographic Society, June 1977.

Encyclopædia Britannica. "Caledonian Canal." Accessed January 3, 2019. www.britannica.com/topic/Caledonian-Canal.

Encyclopedia.com. "Geist, Bill 1945-?" March 11, 2020. Accessed April 16, 2020. www.encyclopedia.com/education/news-wires-white-papers-and-books/geist-bill-1945.

English Heritage. "History of Stonehenge." Accessed December 23, 2019. www.english-heritage.org.uk/visit/places/stonehenge/history-and-stories/history/.

Fort Augustus Abbey. *Fort Augustus Abbey.* Fort Augustus, Scotland: Fort Augustus Abbey, n.d.

Foxboro Foxtrotters. Certificate for completing 1983 Foxtrotter Marathon. Foxboro, MA: Foxboro Foxtrotters, 1983.

Geist, William E. "Village of Lake Champlain Seeking Its Fortune in Tale of a Fabulous Sea Monster." *New York Times,* November 29, 1980.

Gilliland, Haley Cohen. "A brief history of the US Navy's dolphins." *MIT Technology Review*. Cambridge, MA: Massachusetts Institute of Technology, October 24, 2019. Accessed October 25, 2019. www.technologyreview.com/s/614591/dolphin-echolocation-us-navy-war/.

Glengarry Castle Hotel. "Invergarry Castle." Accessed March 22, 2020. www.glengarry.net/castle.php.

Gould, R. T. *The Loch Ness Monster and Others*. London: Geoffrey Bles, 1934.

Gould, Rupert T. *The Loch Ness Monster*. Secaucus, NJ: Citadel Press, 1976 ed.

Greenwell, J. Richard (ed.). "Formation of the Society." *The ISC Newsletter*. Spring, Vol. 1, No.1. Tucson, AZ: International Society of Cryptozoology, 1982.

The Guardian (London). "David James." December 19, 1986.

Hamilton, W. D. and J. Hughes. *The Mysterious "Monster" of Loch Ness*. Fort Augustus, Scotland: Fort Augustus Abbey Press, 1934.

Harmsworth, A. G. "Assessment of the [Loch Ness] Project's Motives and Attitudes." Drumnadrochit, Scotland: A.G. Harmsworth, July 24, 1984.

———. "The Fish and Other Vertebrates in Loch Ness." 2015. Accessed January 23, 2020. www.loch-ness.org/fishandothervertebrates.html#fish.

Harmsworth, Tony. *Loch Ness Monster Explained & Loch Ness Understood*. Drumnadrochit, Scotland: harmsworth.net, 2012.

Healy, Tony and Paul Cropper. *Australian Poltergeist: The Stone-throwing Spook of Humpty Doo and Many Other Cases*. Sydney, Australia: Strange Nation, 2014.

———. *Out of the Shadows: Mystery Animals of Australia*. Sydney, Australia: Ironbark, 1994.

———. *The Yowie: A Search for Australia's Bigfoot*. Sydney, Australia: Strange Nation, 2006.

Heuvelmans, Bernard. *In the Wake of the Sea-Serpents*. New York: Hill and Wang, Inc., 1968. Originally published in French in 1965.

———. *On the Track of Unknown Animals*. New York: Hill and Wang, Inc., 1958. Originally published in French in 1955.

Historic Environment Scotland. "Urquhart Castle." Accessed January 14, 2019. www.historicenvironment.scot/visit-a-place/places/urquhart-castle/history/.

History Is Now Magazine. "Revolutionary War Hero…The Female Paul Revere—Sybil Ludington." Accessed January 8, 2019. www.historyisnowmagazine.com/blog/2017/8/17/revolutionary-war-hero-the-female-paul-revere-sybil-ludington#.XDUfWy2ZMzg=.

Hoare, Callum. "Loch Ness Monster: How 'unidentified' DNA was found in Scottish Highland waters." London: *Daily Express*, June 21, 2020. Accessed January 22, 2020. www.express.co.uk/news/weird/1228931/loch-ness-monster-mystery-solved-dna-found-scottish-highlands-nessie-folklore-spt.

Holiday, F. W. *The Dragon and the Disc*. London: Futura Publications Limited, 1973.

–––. *The Great Orm of Loch Ness*. London: Faber and Faber Limited, 1968.

Holmes, Robin. "The Loch Ness Wellington Revisited." *FlyPast*. pp. 56–59. Stamford, United Kingdom: Key Publishing Ltd., April 1982.

–––. *One of Our Aircraft: The Story of "R for Robert," the Loch Ness Wellington*. London: Quiller Press Ltd. for The Brooklands Museum, 1991.

Hudson, Hugh (director). *Chariots of Fire*. Los Angeles: Twentieth Century-Fox Corporation, Allied Stars Ltd., and Enigma Productions, 1981.

IMDb. *The Private Life of Sherlock Holmes*. Amazon, 1970. Accessed January 14, 2019. www.imdb.com/title/tt0066249/.

James, David. *Escaper's Progress*. London: Corgi Books, 1978.

Johnson, Ben. "Highland Forts of Scotland." Devon, UK: Historic UK. Accessed January 26, 2020. www.historic-uk.com/HistoryMagazine/DestinationsUK/Highland-Forts-of-Scotland/.

Kelbie, Paul. "Underwater marathon man emerges from Loch Ness." *The Independent* (London), October 10, 2003. Accessed January 20, 2020. www.independent.co.uk/news/uk/this-britain/underwater-marathon-man-emerges-from-loch-ness-90707.html.

Kennard, Jim, Roland Stevens, and Roger Pawlowski. *Shipwrecks of Lake Ontario: A Journey of Discovery*. Toledo, OH: Great Lake Historical Society—National Museum of the Great Lakes, 2019.

Klein, Martin. E-mail correspondence with author, 2019.

–––. E-mail correspondence with author, 2020.

–––. "Sonar Search at Loch Ness." *Underwater Search at Loch Ness.* pp. 25–40. Belmont, MA: Academy of Applied Science, 1972.

Klein, Martin and Charles Finkelstein. "Sonar Serendipity in Loch Ness." *Technology Review.* pp. 45–57. Cambridge, MA: Massachusetts Institute of Technology, December 1976.

Kozak, Garry. E-mail correspondence and telephone conversation with author, 2019.

–––. Klein side scan sonar fieldwork at Loch Ness, Scotland, 2005.

Kuenzel, Charles. Conversations with author, 2019.

Lane, W. H. *The Home of the Loch Ness Monster.* Edinburgh: Grant & Murray, 1934.

Leslie, Lionel. *One Man's World: A Story of Strange People and Strange Places.* London: Pall Mall Press, 1961.

Leslie, Lionel A.D., F.R.G.S. *Wilderness Trails in Three Continents—An Account of Travel, Big Game Hunting and Exploration in India, Burma, China, East Africa and Labrador.* London: Heath Cranton Limited, 1931.

Lochnessmystery.blogspot. "Books About the Loch Ness Monster." March 13, 2012. Accessed April 24, 2020. www.lochnessmystery.blogspot.com/2012/03/books-on-loch-ness-monster_13.html.

The Loch Ness Wellington Association Ltd. *The Story of "Another" Loch Ness Monster.* Edinburgh: The Loch Ness Wellington Association Ltd., 1985.

Lopez, Robert. Dr. News Release on Results of 13th Annual Essex County 24 Hour Marathon Relay and Ultramarathon. Westport, NY, July 26, 1987.

Lyons, Stephen. "Famous Photo Falsified?" PBS Nova. Arlington, VA: Public Broadcasting Service, 2000. Accessed April 18, 2020. www.pbs.org/wgbh/nova/lochness/legend3.html.

Mackal, Roy P. *The Monsters of Loch Ness.* Chicago: The Swallow Press Inc., 1976.

–––. *Searching for Hidden Animals—An Inquiry into Zoological Mysteries.* Garden City, NY: Doubleday & Company, Inc., 1980.

Martin, Douglas. "Robert Rines, Inventor and Monster Hunter, Dies at 87." *New York Times.* November 7, 2009.

McKenzie, Steven. "Film's lost Nessie monster prop found in Loch Ness." BBC News (London), April 13, 2016. Accessed December 8, 2019. www.bbc.com/news/uk-scotland-highlands-islands-36024638.

Meaney, Mary Pat. Conversations with author, 2019–2020.

Meredith, Dennis L. *Search at Loch Ness: The Expedition of the* New York Times *and Academy of Applied Science.* New York: Quadrangle/The New York Times Book Company, Inc., 1977.

Mills, J. D. Letter to author, September 24, 1985.

Milne, P. H. "Underwater Stone Circles at Loch Ness." Report provided by author. 1985.

Naone, Erica."The Nessie Quest—Edgerton joined out of friendship and curiosity." *MIT Technology Review.* Cambridge, MA: Massachusetts Institute of Technology, October 15, 2007. Accessed April 29, 2020. www.technologyreview.com/2007/10/15/223418/the-nessie-quest/#.

National Trust for Scotland. "The Battle of Culloden." Edinburgh: National Trust for Scotland. Accessed January 26, 2020. www.nts.org.uk/visit/places/culloden/the-battle-of-culloden.

Naval History and Heritage Command. "USS *Thresher* (SSN-593)." Washington, DC: US Navy. Accessed December 12, 2019. www.history.navy.mil/browse-by-topic/ships/submarines/uss-thresher--ssn-593-.html.

New York Times (New York). "Travel Advisory: Submarine; Loch Ness Trip," April 24, 1994.

Oceaneering International, Inc. "ADS—Atmospheric Diving Suits with Underwater Work Capabilities." Houston, TX: Oceaneering International, Inc., n.d.

The Pace Setter. "14th Annual Essex County 24 Hour Ultra-Marathon." Slingerlands, NY: Hudson-Mohawk Road Runners Club, Month unknown, 1988.

Pepe, Peter and Joseph W. Zarzynski. *Documentary Filmmaking for Archaeologists.* Walnut Creek, CA: Left Coast Press, 2012.

———. "Working Title: 'Sherlock's Missing Loch Ness Movie Monster.'" Glens Falls, NY: Pepe Productions, 2012. Pepe, Peter (director) and Joseph W. Zarzynski (writer). *Wooden Bones: The Sunken Fleet of 1758* (DVD documentary). Glens Falls, NY: Pepe Productions and Bateaux Below, Inc., 2010.

Pepe, Peter (director), Joseph W. Zarzynski (writer), and John Whitesel (writer). *The Lost Radeau: North America's Oldest Intact Warship* (DVD documentary). Glens Falls, NY: Pepe Productions, Bateaux Below, Inc., Whitesel Graphics, and Black Laser Learning, 2005.

Reuters. "Mystery colony 20 to 30 Nessies lurking." *The Ottawa Citizen.* August 12, 1978.

Rines, Dr. Robert. Conversation with author, 1979.

Rock 'n' Roll Marathon Series. "Rock 'n' Roll Marathon Series." Accessed February 18, 2019. www.runrocknroll.com/en.

Ross, David. "John Cobb Memorial, Loch Ness." Accessed February 11, 2020. www.britainexpress.com/scotland/Highlands/properties/cobb-memorial.htm.

Sargent, A. Written statement about a "Lizzie" sighting on September 30, 1975.

Scott, Peter and Robert Rines. "Naming the Loch Ness monster." *Nature.* pp. 466–468. Vol. 258, Issue 5535. London: Nature Research, December 11, 1975.

Searle, Frank. *Around Loch Ness—A Handbook for Nessie Hunters.* Inverness, Scotland: John G. Eccles Ltd., 1977.

Skidmore College. "Skidmore College: 1903–1910, Young Women's Industrial Club." 2020a. Accessed January 25, 2020. www.skidmore.edu/skidmorehistory/1903-1910.php.

———. "Skidmore College: 1965–1986, President Palamountain." 2020b. Accessed January 25, 2020. www.skidmore.edu/skidmorehistory/1965–1986.php.

Smith, Richard D. "Testing an Underwater Video System at Lake Champlain." *Cryptozoology.* Vol. 3, 89-93. Tucson, AZ: International Society of Cryptozoology, 1984.

The Telegraph (London). "The Loch Ness monster eludes Operation Deepscan." October 11, 2016. Accessed April 12, 2020. www.telegraph.co.uk/only-in-britain/operation-deepscan-searches-loch-ness/.

Tikkanen, Amy. "Loch Ness monster." *Encyclopædia Britannica.* Accessed January 17, 2020 and April 18, 2020. www.britannica.com/topic/Loch-Ness-monster-legendary-creature#ref1249155.

The Times (London). "Mr. Lionel Leslie." January 21, 1987.

Trudeau, Garry. "Doonesbury." Kansas City: Universal Press Syndicate, MO, July 19–31, 1976. *UltraRunning* (Bend, OR). 1981-Present.

Valley News (Elizabethtown, NY). "Kouros, Gibbons break Westport track records," July 15, 1987.

Vancouver Sun (Vancouver, BC, Canada). "Loch Ness 'Monster' Sighted," July 26, 1969.

Victor, Daniel. "Loch Ness Monster Is Found! (Kind of. Not Really.)." *New York Times*, April 13, 2016.

Visit Scotland. "Ben Nevis, Scotland's Tallest Peak." 2019a. Accessed December 8, 2019. www.visitscotland.com/see-do/iconic-scotland/ben-nevis/#article-page-1.

———. "Glenfinnan Monument & Visitor Center." 2019b. Accessed January 28, 2019. www.visitscotland.com/info/see-do/glenfinnan-monument-visitor-centre-p246521.

Weir, Tom. *The Scottish Lochs*. Edinburgh: Constable, 1980.

Wilder, Billy (director). *The Private Life of Sherlock Holmes*. Los Angeles: The Mirisch Production Company, 1970.

Wilder, Billy and I.A.L. Diamond. *The Private Life of Sherlock Holmes* (movie script). Accessed May 12, 2020. www.dailyscript.com/scripts/holmes.pdf.

Wilford, John Noble. "Dolphins to Join Loch Ness Hunt." *New York Times*, March 22, 1979.

Witchell, Nicholas. *The Loch Ness Story*. Lavenham, Suffolk: Terence Dalton Limited, 1974.

Whyte, Constance. *More Than a Legend: The Story of the Loch Ness Monster*. London: Hamish Hamilton, 1957.

Zarzynski, Joseph W. "'Champ'—A Zoological Jigsaw Puzzle." *Adirondack Bits 'n Pieces*. Volume 1, Number 1, pp. 16–21. Port Henry, NY: Bannister Publications, 1983.

———. *Champ—Beyond the Legend*. Port Henry, NY: Bannister Publications, 1984 (1st ed.).

———. *Champ—Beyond the Legend*. Wilton, NY: M-Z Information, 1988 (2nd ed.).

–––. *Ghost Fleet Awakened: Lake George's Sunken Bateaux of 1758.* Albany, NY: SUNY Press, 2019.

–––. "LCPI Work at Lake Champlain, 1984." *Cryptozoology.* Vol. 3, pp. 80–83. Tucson, AZ: International Society of Cryptozoology, 1984a.

–––. "LCPI Work at Lake Champlain, 1988." *Cryptozoology.* Vol. 7, pp. 70–77. Tucson, AZ: International Society of Cryptozoology, 1988a.

–––. *Monster Wrecks of Loch Ness and Lake Champlain.* Wilton, NY: M–Z Information, 1986.

–––. Personal journal of July 11–20, 1981 trip to Loch Ness, Scotland, 1981.

–––. Personal journals and notes, 1976–1985.

–––. "Proposal—ISC Morar/Shiel 1983." Wilton, NY: Planning journal, 1983.

–––. "'Seileag': The Unknown Animal(s) of Loch Shiel, Scotland." *Cryptozoology.* Vol. 3, pp. 50–54. Tucson, AZ: International Society of Cryptozoology, 1984b.

Zarzynski, Joseph W. (ed.). *Champ Channels.* Wilton, NY: Lake Champlain Phenomena Investigation, 1983–1989.

–––. "Morag." *Champ Channels.* Vol. 2, No. 1. p. 2. Wilton, NY: Lake Champlain Phenomena Investigation, 1984c.

–––. "Running With Lake Monsters." *Champ Channels.* Vol. 2, No. 3. p. 1. Wilton, NY: Lake Champlain Phenomena Investigation, 1984d.

Zarzynski, Joseph W. and Bob Benway. *Lake George Shipwrecks and Sunken History.* Charleston, SC: The History Press, 2011.

Zarzynski, Joseph W. and M. Pat Meaney. "Investigations at Loch Ness and Seven Other Freshwater Scottish Lakes." *Cryptozoology.* Vol. 1, Winter. pp. 78–82. Tucson, AZ: International Society of Cryptozoology, 1982.

Index

About the Author

Joseph W. Zarzynski is a resident of Saratoga County, New York. He taught social studies for 31 years in the Saratoga Springs City School District. His career also included years as a cryptozoologist (monster hunter), underwater archaeologist, author, newspaper columnist, and documentary scriptwriter. He has a Bachelor of Arts degree (History, 1973) from Ithaca College, a Master of Arts in Teaching degree (Social Sciences, 1975) from Binghamton University, and a Master of Arts degree (Archaeology and Heritage, 2001) from the University of Leicester, UK. From 1974 to 1991, Zarzynski conducted numerous cryptozoological expeditions at Loch Ness,

Scotland and at "North America's Loch Ness"—Lake Champlain.

A self-described "average" marathoner and ultramarathoner, in 1984, he completed a solo 28.5-mile run along the full length of Loch Ness. He is believed to be the first person to have accomplished that ultrarun. In 1985, as a correspondent for *General Aviation News*, Zarzynski reported on the recovery of a sunken World War II British Wellington bomber from Loch Ness. From 1987 to 2011, the Endicott, New York native was executive director of Bateaux Below. For a quarter-of-a-century, the not-for-profit archaeological team studied shipwrecks in Lake George, New York, primarily French & Indian War (1755–1763) vessels. In 1990, Zarzynski directed the group that used a Klein side scan sonar to discover Lake George's 1758 *Land Tortoise* radeau shipwreck, called "North America's Oldest Intact Warship." The author has collaborated on four shipwreck documentaries produced by Pepe Productions. He's had published over

175

500 newspaper stories, newsletter articles, professional journal papers, and scientific reports. Zarzynski is author or co-author of seven books, four on maritime archaeology topics and three about underwater mysteries. He enjoys jogging, writing, and with his wife, volunteering on archaeological excavations and cataloguing artifacts for museums.